# Eleventh International Conference on Atomic and Molecular Data and Their Applications

# Eleventh International Conference on Atomic and Molecular Data and Their Applications

Special Issue Editors

**James F. Babb**
**Nancy Brickhouse**

MDPI • Basel • Beijing • Wuhan • Barcelona • Belgrade

*Special Issue Editors*
James F. Babb
Center for Astrophysics |
Harvard & Smithsonian
USA

Nancy Brickhouse
Center for Astrophysics |
Harvard & Smithsonian
USA

*Editorial Office*
MDPI
St. Alban-Anlage 66
4052 Basel, Switzerland

This is a reprint of articles from the Special Issue published online in the open access journal *Atoms* (ISSN 2218-2004) in 2019 (available at: https://www.mdpi.com/journal/atoms/special_issues/Atomic-MolecularData).

For citation purposes, cite each article independently as indicated on the article page online and as indicated below:

LastName, A.A.; LastName, B.B.; LastName, C.C. Article Title. *Journal Name* **Year**, *Article Number*, Page Range.

**ISBN 978-3-03928-088-9 (Pbk)**
**ISBN 978-3-03928-089-6 (PDF)**

# Contents

# About the Special Issue Editors

**James Babb** (PhD) is a physicist at the Center for Astrophysics | Harvard & Smithsonian, a Fellow of the American Physical Society, a member of the American Astronomical Society, and a member of the scientific board of the Institute for Theoretical Atomic, Molecular, and Optical Physics (ITAMP). He has been active in applications of atomic and molecular data since the exploratory ITAMP workshop "Atomic and Molecular Data for Science and Technology" (June 1996, Cambridge, MA), and has served on the Conference Committees of the first, eighth, and eleventh ICAMDATA meetings (1997, 2012, 2019) and as Secretary (2004-8, 2017-18) and Member (2000-4, 2019-present) of the ICAMDATA International Program Committee. Dr. Babb is the Founding Editor-in-Chief of *Atoms*.

**Nancy Brickhouse** (PhD) is Senior Science Advisor at the Center for Astrophysics | Harvard & Smithsonian. She served as Associate Director for the Solar, Stellar, and Planetary Sciences Division from 2004 to 2012. She was elected to the American Astronomical Society Council for the term 2012–2015. Her research interests include solar and stellar coronal physics, plasma spectral modeling, atomic data for astrophysics, UV to X-ray spectroscopy of diverse objects, and physical processes in astrophysical plasmas. She is Leader of the ATOMDB Project, which uses collisional and radiative atomic data to generate spectral models needed for high-energy astrophysics. Dr. Brickhouse received her BS in Physics from the University of North Carolina at Chapel Hill, and her PhD in Physics from the University of Wisconsin–Madison.

# Preface to "Eleventh International Conference on Atomic and Molecular Data and Their Applications"

**ICAMDATA 2018**

**The 11th International Conference on Atomic and Molecular Data and Their Applications**

The 11th International Conference on Atomic and Molecular Data and their Applications (ICAMDATA) was held on November 11–15, 2018, in Cambridge, Massachusetts, and was organized by the Center for Astrophysics | Harvard & Smithsonian. This meeting is a continuation of a series which began in 1997 that was chartered to promote the use of atomic and molecular (AM) data in various fields of science and technology, to provide a forum for the interaction of AM data producers and users, and to foster crossdisciplinary cooperation between AM data producers and users as the coordination of AM data activities and databases worldwide.

The conference was held at the Sheraton Commander Hotel, located adjacent to the historic Cambridge Common. A welcome reception was held at the Center for Astrophysics on Sunday evening. The conference was opened on Monday November 12, and continued until Thursday with sessions of invited and contributed talks. During the meeting, there were two sessions for poster viewing, a panel discussion on trends in data publication, and two sessions dedicated to database demonstrations. The conference dinner was held on Wednesday evening at the Harvard Faculty Club.

There were 60 participants representing 15 countries that attended the conference: 18 invited talks, 7 database demonstrations, and 33 contributed talks and poster papers were presented.

We are grateful to Margaret Carroll, Joyce Silberman, and Debbie Nickerson from the Director's Office of the CfA for their tireless and enthusiastic assistance in organizing the conference. Further local assistance from Naomi Tariri, Rob Hargreaves, and Mike Capone is acknowledged. We thank Gordon Drake for contributing conference photos. Finally, incoming International Program Secretary Iouli Gordon helped in all phases as the third member of the Local Conference Committee.

We are grateful to Charles Alcock for partial meeting support from the CfA Director's Office. We also thank the Institute for Theoretical Atomic, Molecular, and Optical Physics (ITAMP), supported by the NSF, which provided partial funding for several international speakers.

At the conclusion of the conference, Annarita Laricchiuta became Chair of the International Program Committee and the next ICAMDATA conference will be held in Bari, Italy.

James F. Babb

Nancy Brickhouse

**ICAMDATA 2018 Local Conference Committee**
Nancy Brickhouse (Chair), James Babb, Iouli Gordon

**ICAMDATA 2018 International Program Committee**
James Babb, Acting Secretary (USA)
Beatriz Barbuy (Brazil)
Jacek Bieron (Poland)
Igor Bray (Australia)
Nancy Brickhouse, Chair (USA)
Hyuck Cho, Past Local Chair (Korea)
James Colgan (USA)
Chenzhong Dong (China)
Gordon Drake (Canada)
Stephan Fritzsche (Germany)
Iouli Gordon, Incoming Secretary (USA)
Christian Hill (IAEA)
Jelle Kaastra (Netherlands)
Annarita Laricchiuta, Vice-Chair (Italy)
Izumi Murakami (Japan)
Nobuyuki Nakamura (Japan)
Martin O'Mullane (UK)
Gerry O'Sullivan (Ireland)
Juliet Pickering (UK)
Yuri Ralchenko (USA)
Stefan Schippers (Germany)
Viatcheslav Shevelko (Russia)
Jianguo Wang (China)
Jung-Sik Yoon, Past Chair (Korea)

**Complete List of Published ICAMDATA Proceedings and Collections**

1. 1997. Gaithersburg, Maryland. AIP Conf. Proc. 434, Peter J. Mohr and Wolfgang L. Wiese, eds. (AIP, Woodbury, NY, 1998); poster papers in NIST Special Pub. 926, Wolfgang L. Wiese and Peter J. Mohr, eds. (US GPO, Washington, DC, 1998).
2. 2000. Oxford, England. AIP Conf. Proc. 543, Keith A. Berrington and Kenneth Bell, eds. (AIP, Melville, NY, 2000).
3. 2002. Gatlinburg, Tennessee. AIP Conf. Proc. 636, David R. Schultz, Predrag S. Krstic and Fay Ownby, eds. (AIP, Melville, NY, 2002).
4. 2004. Toki, Japan. AIP Conf. Proc. 771, Takako Kato, Hisamichi Funaba and Daiji Kato, eds. (AIP, Melville, NY, 2005); contributed papers in J. Plasma Fusion Res. Ser. (JPFR-S) 7 (2006), T. Kato, H. Funaba, C. Suzuki, T. Minami, eds.; meeting summary by R. K. Janev and T. Kato, Phys. Scr. 72, C6 (2005).
5. 2006. Meudon, France. AIP Conf. Proc. 901, Evelyne Roueff, ed. (AIP, Melville, NY, 2007).
6. 2008. Beijing, China. AIP Conf. Proc. 1125, Shaoping Zhu and Jun Yan, eds. (AIP, Melville, NY, 2009).
7. 2010. Vilnius, Lithuania. AIP Conf. Proc. 1344, A. Bernotas, R. Karazija, and Z. Rudzikas, eds. (AIP, Melville, NY, 2011).
8. 2012. Gaithersburg, Maryland. AIP Conf. Proc. 1545, John D. Gillapsy, Wolfgang L. Wiese, and Yuri A. Podpaly, eds. (AIP, Melville, NY, 2013).
9. 2014. Jena, Germany. Phys. Scr. 90 (5), 050201 (2015), Stefan Schippers, Andrey Surzhykov, and Stephan Fritzsche, eds.
10. 2016. Gunsan, South Korea. Eur. Phys. J. D 72, 49 (2018), Gordon W. F. Drake, Jung-Sik Yoon, Daiji Kato, and Grzegorz Karwasz, eds.
11. 2018. Cambridge, Massachusetts. Atoms 7 (2019), James F. Babb and Nancy Brickhouse, eds.; contributed abstracts at https://projects.iq.harvard.edu/files/icamdata/files/icamdata-abs-final.pdf

**James F. Babb, Nancy Brickhouse**
*Special Issue Editors*

atoms

MDPI

Article

# Cowan Code: 50 Years of Growing Impact on Atomic Physics

**Alexander Kramida**

National Institute of Standards and Technology, Gaithersburg, MD 20899, USA; alexander.kramida@nist.gov

Received: 28 May 2019; Accepted: 19 June 2019; Published: 2 July 2019

**Abstract:** The famous Cowan's book, "The Theory of Atomic Structure and Spectra", published in 1981, and his suite of computer codes based on it, continue to be highly influential in atomic physics and many other research areas. As of September 2018, there have been more than 5000 citations to Cowan's book and codes, and each year adds about 150 citations to this list. The present work briefly describes what these codes do and why they are responsible for most of the current progress in the analyses of atomic spectra. Various modifications of these codes, including my own, will also be described.

**Keywords:** Cowan code; atomic structure; energy levels; transition probabilities; semiempirical parametric calculations

## 1. Introduction

Although Cowan's classic book [1] was published in 1981, the early versions of his computer codes that are generally referred to this book were fully developed as early as in 1968 [2], which accounts for the 50-year period mentioned in the title of the present paper. The program package usually called "the Cowan code" consists of four separate codes. Some parts of this package were developed even earlier [3,4].

When I received an invitation to make a talk at the 11th International Conference on Atomic and Molecular Data and their Applications (ICAMDATA-2018), Robert Duane Cowan was still alive (see Figure 1). He died on 26 July 2018 in a hospice in Albuquerque at an age of 98. The main facts of his biography can be found in an obituary available online [5]. The present article is a transcript of my talk given on 13 November 2018 at Cambridge, Massachusetts.

**Figure 1.** Robert Duane Cowan (24 November 1919–26 July 2018).

## 2. Impact of Cowan's Book and Computer Codes

According to the citation index of the Institute for Scientific Information (ISI) Web of Science database [6], there have been more than 5000 citations of Cowan's book [1] and the codes based on it. The citing articles cover a wide range of 58 research categories, ranging from optics, atomic physics, solid state physics, astronomy, and chemistry to nuclear physics (see Figure 2). Figure 3 shows the dynamics of these citations from 1981 to 2017. Since 1990, there have been about 150 new citations to Cowan's work every year. This shows that Cowan's legacy continues to have a great influence on science.

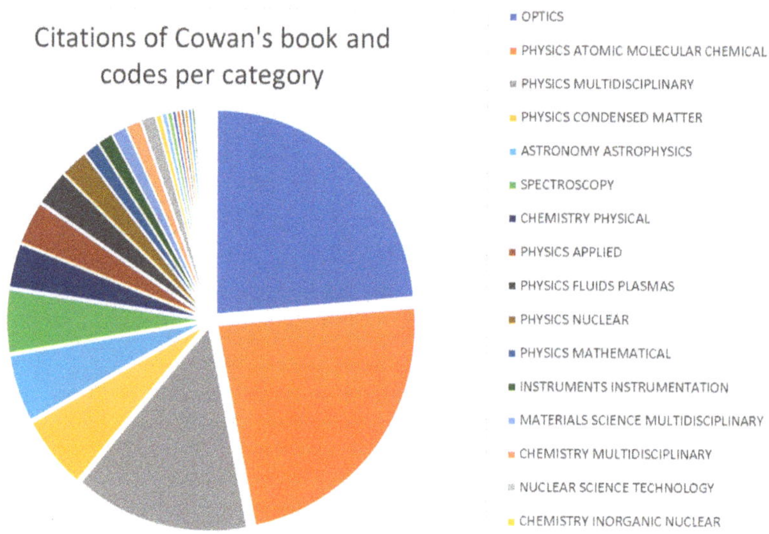


**Figure 2.** Distribution of 5026 citations to Cowan's book [1] and the computer codes based on it, per research category (per Cited Reference Search of Web of Science [6]).

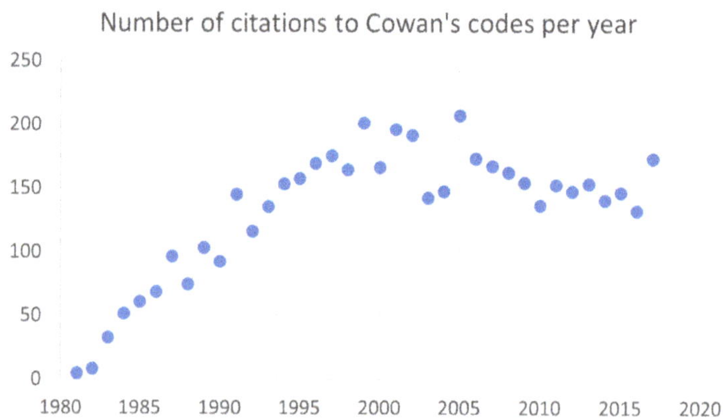


**Figure 3.** Number of citations to Cowan's book [1] and computer codes based on it per year (per the same source as Figure 2).

In the following sections, I will try to explain why Cowan's book and codes are so important to modern atomic physics. I will briefly describe the predecessors of Cowan's codes, explain what those codes do, enumerate other existing atomic physics codes, including various versions of Cowan's codes, point out some known problems in these codes, and outline the needs for future development.

## 3. What Cowan's Codes Do

The package of Cowan's codes consists of four Fortran programs. The names of these programs are made of Cowan's initials, 'RC', followed by a code letter. Some code names are followed by a part or version number. The codes are intended to be run in sequence, and each program produces output files that are simultaneously input files for the next code in the chain.

The calculations start with the code RCN, which calculates single-configuration radial wavefunctions for a spherically symmetrized atom via the Hartree–Fock method. Next, RCN2 calculates the radial integrals (Slater parameters), including the configuration interactions (all terminology is described in [1]). Then, RCG, which is the main code of the package, computes the angular matrix elements of the Hamiltonian, and by diagonalizing it, computes the energy levels, transition wavelengths, and radiative rates for the electric-dipole (E1), magnetic-dipole (M1), and electric-quadrupole (E2) transitions. RCG can also compute autoionization rates and plane-wave Born excitation cross-sections for electronic collisions. Finally, RCE does a least-squares fitting of atomic energy levels (i.e., adjusts the Slater parameters to fit experimental levels). The adjusted Slater parameters can be transferred to the input file of RCG. Then, re-running RCG produces much more accurate transition rates. The extent of improvement depends on the quality of the least squares fit and can be as large as orders of magnitude for some transitions. However, one can rarely expect the accuracy of the calculated transition rates to be better than 20% for the strongest transitions, due to the limitations of the method. No general estimate of uncertainties can be made. These uncertainties must be analyzed by comparisons with other data and evaluated for each calculation.

## 4. Cowan's Predecessors

Before Cowan's codes were developed, the basic theory was already known, and there were other computer codes developed in early 1960s. The most well-known of them was the suite of the so-called French codes (AGENAC, ASSAC, DIAGAC, and GRAMAC). They were created in Laboratoire Aimé Cotton in Orsay near Paris. In these acronyms, the last two letters "AC" stand for "Aimé Cotton." The French codes did most of the same things as Cowan's in similar steps:

- Compute the wavefunctions.
- Compute the Slater parameters. It was possible to also compute parameters of additional interactions, such as hyperfine. The user had to construct the input files from the output of the previous code in a laborious semi-manual procedure.
- Assemble the Hamiltonian matrix (extremely tenuous semi-manual task). It was flexible, allowing for introduction of additional interactions.
- Diagonalize the Hamiltonian.
- Use the output of the diagonalization to compute the atomic structure. Calculation of transition rates was not included in the codes. The users had to build their own code extensions to do that.
- Use a least-squares fitting code to adjust the Slater parameters.

All steps involved the difficult manual work of constructing the input files and parsing the output files, which made computations very inefficient. However, the ability of calculating hyperfine structure, isotope shifts, and effects of other interactions (such as external electric and magnetic fields or weak interactions neglected in Cowan's codes) made the French suite of codes indispensable. It is quite unfortunate that these codes no longer exist. The source codes have never been published. The same is true for Cowan's suite of codes. However, Cowan's codes have been preserved in the Los Alamos National Laboratory, where he worked, and are available online (see Section 7).

A copy of the manual for the French suite of code still exists in the archives of the Atomic Spectroscopy Group of the National Institute of Standards and Technology (NIST). The first page of this manual is shown in Figure 4.

**CHAINE DE PROGRAMMES**

**PROVISOIRE**

**AGENAC, ASSAC 2, DIAGAC, GRAMAC**

En attendant une chaine de programmes complètement rénovée, une chaine provisoire a été constituée; tous les programmes de cette chaine sont compatibles, sur bande et sur cartes; ils sont compilés en Fortran V et ce sont eux qui seront désormais sur Fastrand.

Les anciens programmes ne sont pas pour autant détruits et seront utilisables sur bande en cas d'incidents. Les anciens résultats d'Agenac ne sont pas compatibles avec le nouvel Assac, par contre les configurations des bibliothèques sont toujours compatibles.

On trouvera ci-dessous un mode d'emploi de chacun de ces programmes ( les remises à jour éventuelles seront distribuées). On trouvera aussi la valeur des paramètres de chaque programme : si un cas particulier se présentait, consulter :

- pour AGENAC : Annick CARLIER
- pour ASSAC 2 : Yves BORDARIER
- pour DIAGAC : Pierre DAGOURY
- pour GRMC 1 : Yves BORDARIER ou Pierre DAGOURY.

Suivant les cas, on pourra obtenir une version particulière à utiliser soit sur bande, soit sur Fastrand, ou bien on sera prié de modifier soi-même un exemplaire du programme mis à sa disposition.

Par contre, les erreurs doivent être signalées immédiatement au responsable ci-dessus ou à défaut à Yves BORDARIER et seront corrigées dans les plus brefs délais.

**Figure 4.** The first page of the manual for the French suite of atomic codes.

Very little is known about the authors of these codes. Yves Bordarier (1930–2014) was a brilliant theoretician. His Ph.D. thesis was devoted to the procedures of summation for Wigner's $6j$-symbols. His solutions found applications not only in atomic physics, but also in crystallography. In collaboration with other well-known atomic physicists, such as J. Blaise, B. R. Judd, M. Klapisch, and G. G. Gluck, he published three papers on the hyperfine structure of Tm [7] and Eu [8], and on isotope shifts and the electronic structure of Os atoms [9]. Annick Carlier also worked on the interpretation of electronic structure and isotope shifts in lanthanide atoms, such as Sm [10]. Little is known about Pierre Dagoury. He was a computer engineer and worked in the Aimé Cotton Lab for only a short time. These people were brilliant scientists. The codes they created in this early time of computer technology allowed successful analyses of such complex lanthanide spectra that even today are extremely difficult to analyze.

## 5. The Workflow of Cowan's Suite of Codes

Most of the calculational procedures of Cowan's codes look similar to those of the French suite:

- Specify the atomic configurations and code options in two input files.
- Run RCN (produces input for RCN2).
- Run RCN2 (produces input for RCG).
- Run RCG (cringe at the results, as they are rather inaccurate). This produces the input for RCE and outputs nicely formatted transition data.
- Run RCE. To fit experimental levels, edit an input file and rerun RCE.
- Use utility codes to transfer fitted parameters to the RCG input file.
- Re-run RCG and smile! The results are now very accurate.

One difference from the French suite is that all the tasks are automated (although manual intervention, such as changing the program options, is still allowed), except for the least-squares fitting (LSF) with RCE. Running the LSF still involves a lot of manual work and requires much expertise acquired by trial and error. There is no strict methodology for it; some even call it 'black magic' [11]. However, some general recipes do exist, and by knowing them, the analysis of an atomic spectrum with Cowan's codes takes from a few hours to a few weeks, depending on the complexity of the problem. This is much more efficient than working with the French codes, which typically required years for an analysis of one spectrum. The downside of Cowan's codes is their limited flexibility, where they do not allow for the analysis of hyperfine structure and isotope shifts.

Another important difference is the ability of Cowan's codes to calculate transition probabilities. This enables comparisons between predicted and experimental intensities of spectral lines, which increases the reliability of line identifications and sometimes allows a unique choice between multiple possible level assignments. In the case of Tm II, investigated by Wyart [12], such comparisons led to the revision of previous wrongly assigned quantum numbers $J$ of many energy levels.

In the calculational workflow shown above, the statement about the improvement of accuracy by the LSF is, of course, only qualitative. The quality of the fit depends crucially on the inclusion of important interacting configurations in the calculation, as well as the choice of free and constrained parameters. While the accuracy of the fitted and predicted energies can be assessed easily in a straightforward way (by comparison of calculated and experimental levels), uncertainties of the predicted transition rates are much more difficult to estimate. The methodology for such estimation has been developed at NIST and is described in my review [13].

## 6. Why are Cowan's Codes So Important?

From the point of view of modern atomic physics, Cowan's codes are rather primitive. They use the non-relativistic Hartree–Fock method, with only some of the relativistic effects accounted for as perturbations. Moreover, the radial wavefunctions are computed in a hydrogenic single-configuration approximation, and they are frozen in calculations of configuration interactions. That is why Cowan's method is usually called a superposition of configurations, rather than a multiconfiguration calculation. The latter, unlike Cowan's RCN, varies the radial wavefunctions of each basis configuration in the self-consistent field calculation. Nowadays, a number of much more sophisticated multiconfiguration codes exist:

- MCHF [14], CIV3 [15], ... (non-relativistic with relativistic corrections)
- MCDHF [16], MCDFGME [17], FAC [18], MR-MP [19], ... (relativistic)

A more extensive, although still incomplete, list of modern ab initio codes can be found in Section 9.2. These ab initio codes can provide better accuracy, but only for a limited number of spectra, such as those isoelectronic to H, He, Ne, Ar, Kr, K, Rb, Cs, F, Cl, Br, Be, Mg, and Ca (to some extent), those isoelectronic to B, Al, Ga, C, Si, Ge, N, P, As, O, S, Se, Ni, Pd, Pt, Cu, Ag, and Au (to a much lesser extent), and only for a limited number of low-excitation levels. However, for $n$d- and $n$f-transition elements, these codes are equally or even less accurate than Cowan's code. For heavy elements such as lanthanides and actinides, they even fail to reproduce the experimentally known ground states (meaning that they predict incorrect levels or configurations to be the lowest ones in energy). This reflects the fact that ab initio multiconfiguration methods cannot handle atomic systems where electron-correlation effects (i.e., configuration interactions, CIs) are large and numerous. This problem is illustrated in Figure 5, depicting a Grotrian diagram of the Fe II spectrum [20]. Strong CIs occur between closely-lying levels of the same parity and the same *J* values. One can see that as the energies approach the ionization limit, the density of levels belonging to various Rydberg series crossing each other increases, leading to an increasing number of strong configuration interactions.

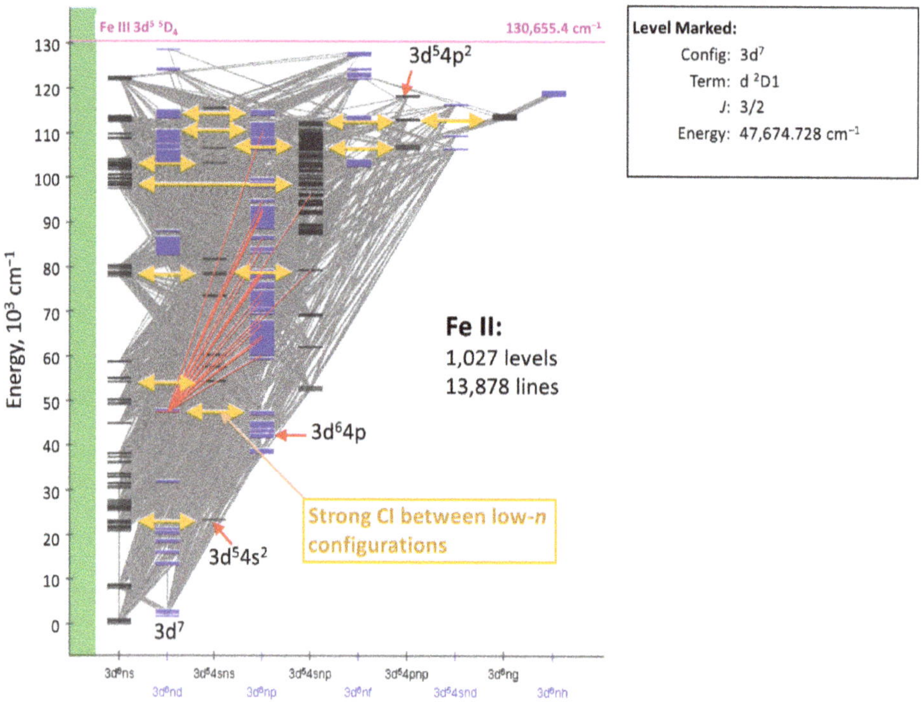


**Figure 5.** Grotrian diagram of the Fe II spectrum [20]. The yellow horizontal double-arrows indicate strong configuration interactions.

An exact account of the CI effects requires a precise determination of level energies. One should recall that the multiconfiguration methods try to achieve it via the addition of a large number of virtual excitations to high-lying configurations. Thus, the precise determination of positions of highly-excited Rydberg levels requires a huge number of configurations, growing exponentially with increasing excitation energy. Even for such a moderately heavy atom as Fe, the dimension of the required matrices becomes unmanageably large (even for supercomputers) for levels that are higher than about half of the ionization energy, i.e., for most of experimentally known levels. For heavier atoms, such as lanthanides and actinides, the number of interacting Rydberg series is much greater, and the energy region riddled by strong CI effects descends to the vicinity of the ground level, making the entire energy structure unmanageable for computation with multiconfiguration methods.

On the other hand, with Cowan's codes, the LSF of all levels, with a standard deviation $<100\ cm^{-1}$, can be accomplished in one week, reproducing all strong line intensities and predicting unknown levels with a similar accuracy. Thus, for the spectra of transition elements, including lanthanides and actinides, analysis using LSF with Cowan's codes remains the only working tool for the past few decades and is expected to remain such for a few more decades.

## 7. Versions of Cowan's Codes

The latest mainstream version of Cowan's code package is presently distributed by Cormac McGuinness at the University of Dublin, Ireland [21]. This package contains Cowan's original source codes and documentation in addition to installation routines written for Linux operating systems.

My version, originally developed for Windows-based personal computers, had branched from the mainstream version circa 1989 and has mutated quite substantially since then. It is now distributed from a NIST website [22]. This package also includes the source codes and documentation, as well as several utility codes. It will be described in more detail in the next section.

There are also many other private or proprietary versions of Cowan's codes. The ones I am aware of are listed below:

- P. Quinet's version with core polarization (private) [23].

  The only codes modified in this version are RCN and RCN2.

- R. L. Kurucz's version (private).

  Branched from mainstream circa 1970. Designed for very large configuration sets and uses large workstations. The famous Kurucz's Atoms collection of computed atomic line lists [24] was produced with these codes.

- J. Ruczkowski, M. Elantkowska, and J. Dembczyński's version (private) [25].

  Incorporates LSF of transition matrix elements in addition to Slater integrals.

- Los Alamos CATS (Cowan ATomic Structure) code (proprietary) [26].

  Parallelized, with dynamic memory allocation, for a large computer cluster. This version does not have an LSF code.

## 8. My Version of Cowan's Codes

I started using Cowan's codes in 1981, when I worked at the Institute of Spectroscopy in Troitsk, Russia. The only available computer that could run this code was a Russian adaptation of PDP-11. Cowan's original codes were adapted for it by Dr. Dmitri S. Viktorov. One peculiar feature of that computer was that its processor held upper-case Cyrillic character symbols in the memory region that was originally intended to contain lower-case Latin letters. Thus, no lower-case Latin letters were available in this system, and all source codes and input/output files had to be converted to upper case. This is the historical reason for upper-case characters appearing in the source codes and some of the input files of my version of Cowan's codes.

I started changing the code near 1989. My starting point was a version modified (after Viktorov) by Yu. Ralchenko to produce level designations using seniority quantum numbers instead of the Nielson and Koster indexes used in Cowan's original RCG code. Since then, I have ported these codes to several different compilers working under Microsoft Windows, VAX-VMS, and several versions of Unix, including Linux and MacOS. Versions prior to 1994 did not have any major differences from the mainstream. In 1994–1995, when I was a guest researcher at NIST, I tried to calculate the spectra of some lanthanide atoms and found that the space needed to store the temporary files required by RCG exceeds the capacity of hard drives typically available then in personal computers. To alleviate the problem, it was necessary to compress the sparse matrices stored in those temporary files. The implementation of this compression was the first major change I made to the codes. At about the same time, Craig Sansonetti and I independently discovered an overflow bug in the term-sorting routine of RCG, which led to the first major bug fix that I made. Since then, I have made a large number of calculations in all isoelectronic sequences from H I to Ds I (hydrogen to darmstadtium), which inevitably revealed numerous other bugs that were lying dormant in the codes.

Besides fixing the bugs, in 1999, I made a drastic modification to the LSF code RCE. In Cowan's original implementation, the input file for RCE is created by RCG and consists of several sections that have no information about eigenvector compositions of energy levels, which makes the association of theoretical and experimental levels a time-consuming and error-prone work. Furthermore, the association of theoretical and experimental levels between the LSF iterations could be made using only two options: Sorting of eigenvalues in increasing order of energies in each $J$+parity group or sorting by calculated Landé factors. Both options perform poorly, making the iterations unstable. In my version, a different algorithm, based on eigenvector recognition [13] was implemented. This implementation includes an alternative input file for RCE, which is created by RCE itself on its first run. This file has a much easier-to-use layout, where each eigenvalue is accompanied by its eigenvector, which can easily be identified with experimental level classifications, and each Slater parameter is immediately followed by its controlling options. Working with this file made the LSF process much more efficient and the convergence of the iterations more stable.

Additionally, my version of Cowan's code package contains a number of utility codes for various manipulations of the input/output files. The most widely used are the following:

- 'printout': Converts the RCE output to nicely formatted tables of energy levels (including the intermediate quantum numbers omitted in Cowan's original output) and LSF parameters.
- 'conv_out': Converts the RCG output to nicely formatted tables of levels and lines. This code can also create input files for the visual line-identification code IDEN2 [27].
- 'update11': Transfers fitted LSF parameters from the RCE output to the RCG input file, re-runs RCG, and runs 'printout'.
- 'reorder_ing11': Converts the RCG input file to change the order of shell summation.

Instructions for the usage of all these codes are given in the writeup provided with the package and are also available from the codes (they are printed if the codes are executed without parameters).

It should be noted that codes similar to IDEN2 [27], mentioned above, which allow handling and visualizing a great amount of calculated data from Cowan's code together with experimental data, mainly account for the progress in increasing the speed of spectral analyses of complex spectra in the last few decades.

## 9. Other Codes with Similar or Better Capabilities

### 9.1. Semiempirical Codes with Parametric Fitting

A powerful orthogonal operators code was developed by Raassen and Uylings [28] in the Zeeman Laboratory, Amsterdam, Netherlands. It implements the concept of orthogonal operators, described by Hansen et al. [11]. Although this code is similar to Cowan's, in that it uses a non-relativistic approximation with relativistic corrections accounted as perturbations, it additionally accounts for small second-order interactions, which are omitted in Cowan's codes. The methodology of orthogonal operators is qualitatively very different from Cowan's parametric calculations. Its capabilities are illustrated in Table 1, which is an abbreviated version of Table 1 by van het Hof et al. [29] (the original table contains also references to the analyses of each spectrum).

Table 1 shows in bold font the standard deviation $\sigma$ obtained in the LSF for each investigated spectrum of $3d^n4s$ configurations. This parameter indicates the accuracy with which experimentally unknown levels can be predicted by LSF. It can be seen that in all cases, $\sigma$ is of the order of 1 $cm^{-1}$, or even smaller. This is really a spectroscopic accuracy for these spectra, and it is two orders of magnitude smaller than the uncertainty achievable with Cowan-code LSF.

The power of the orthogonal operators method was recently demonstrated by the successful analyses of the Re III and Os III spectra by Azarov and Gayasov [30] and Azarov et al. [31]. The last previous attempt to analyze these spectra was made in 1993 by Wyart et al. [32], who deemed it hopeless, despite the large astronomical interest in these spectra.

**Table 1.** Survey of the $3d^n4s$ configurations investigated using orthogonal operators.

| Conf.: | 3d4s | $3d^24s$ | $3d^34s$ | $3d^44s$ | $3d^54s$ | $3d^64s$ | $3d^74s$ | $3d^84s$ | $3d^94s$ |
|---|---|---|---|---|---|---|---|---|---|
| $N_c$: | 4 | 16 | 38 | 63 | 74 | 63 | 38 | 16 | 4 |
| $N_0$: | 5 | 17 | 30 | 42 | 42 | 42 | 30 | 17 | 5 |
| Spectrum: | Ti III | V III | Cr III | Mn III | Fe III | Co III | Ni III | Cu III | Zn III |
| $N_e$: | 4 | 16 | 37 | 54 | 65 | 58 | 38 | 16 | 4 |
| **$\sigma$ [cm$^{-1}$]:** | **0** | **0.4** | **0.6** | **1.4** | **0.8** | **1.8** | **0.7** | **0.6** | **0** |
| Spectrum: | V IV | Cr IV | Mn IV | Fe IV | Co IV | Ni IV | Cu IV | Zn IV | Ga IV |
| $N_e$: | 4 | 16 | 17 | 62 | 59 | 51 | 37 | 16 | 4 |
| **$\sigma$ [cm$^{-1}$]:** | **0** | **0.2** | **1.0** | **1.0** | **1.2** | **1.8** | **0.7** | **0.7** | **0** |
| Spectrum: | Cr V | Mn V | Fe V | Co V | Ni V | Cu V | Zn V | Ga V | Ge V |
| $N_e$: | 4 | 16 | 36 | 57 | 68 | 47 | 37 | 15 | 4 |
| **$\sigma$ [cm$^{-1}$]:** | **0** | **0.4** | **0.4** | **1** | **0.7** | **1.4** | **0.9** | **0.5** | **0** |
| Spectrum: | Mn VI | Fe VI | Co VI | Ni VI | | | | | As VI |
| $N_e$: | 4 | 15 | 36 | 57 | | | | | 4 |
| **$\sigma$ [cm$^{-1}$]:** | **0** | **0.3** | **0.6** | **1.2** | | | | | **0** |

$N_c$: Total number of levels in the configuration. $N_0$: Total number of parameters used in the least-squares fitting (LSF). Some of these parameters were fixed at the calculated values. $N_e$: Number of established levels in the configuration.

The method of orthogonal operators was developed in the Zeeman Laboratory over the course of several decades. Dismantling of the atomic spectroscopy research group in 1999 had put an unfortunate stop to this development. Since then, Ton Raassen worked in the Institute for Space Research of the Space Research Organization of the Netherlands (SRON), while Peter Uylings was teaching physics courses at the University of Amsterdam. After their official retirement this year, Raassen and Uylings have announced that they are returning to atomic physics research and renewing the development of the orthogonal operators code [33].

Up until now, the orthogonal operators formalism has been implemented only for the $d^n$, $d^nl$, $d^nl_1l_2$, and $f^2$ configurations. For other configurations with partially filled f shells, the theory is yet to be developed.

There exist other atomic codes capable of making the LSF with similar or better accuracy than Cowan's codes. One such code is that of Ruczkowski, Elantkowska, and Dembczyński [25], which is a modification of Cowan's codes, incorporating LSF of transition matrix elements in addition to Slater integrals.

Another important extension of Cowan's method was developed by J.-F. Wyart at the Aimé Cotton Laboratory in Orsay, France. It is called a generalized least-squares fit (GLS) and is based on the empirical fact that the Slater parameters behave smoothly along sequences of spectra that are similar in some sense. For example, Blaise et al. [34] employed the GLS method to analyze the $5f^n$ and $5f^n7s$ configurations in a sequence of the first spectra of actinides having different occupation numbers $n$. In this sequence, the Slater parameters $P_k$ were found to vary smoothly as a function of $n$:

$$P_k = A(P_k) + (n-7)B(P_k) + (n-7)^2C(P_k), \tag{1}$$

where empirical constants $A$, $B$, and $C$ can be found by adding a set of equations similar to Equation (1) to the equations expressing the deviations of calculated levels from experimental ones and solving this extended system of equations in a least-squares procedure.

Similarly, Wyart et al. [35] expressed the smooth variation of Slater parameters along the platinum isoelectronic sequence Au II–Bi VI as

$$P_k = A(P_k) + B(P_k)Z_c + C(P_k)/[Z_c + D(P_k)], \tag{2}$$

where $Z_c$ is the core charge. They used Equation (2) in the GLS. Introducing equations similar to Equations (1) and (2) in the least-squares fit of several spectra makes the fitting more stable and dependable, since it reduces the ratio of the number of free parameters to the number of experimentally known energy levels.

The GLS code was written by J.-F. Wyart as an extension to the French codes described earlier (it makes use of the output from DIAGAC and GRAMAC). This code has not been used since the mid-1990s and is effectively dead nowadays, as the French codes went out of use. However, its basic concepts are fairly simple, and it should not be very difficult to rewrite this code for modern computer platforms, to which Cowan's codes have already been ported.

### 9.2. Other Ab Initio Codes

As mentioned in Section 6, there exist a number of powerful ab initio codes that are similarly or more accurate than Cowan's code (in its ab initio or scaled Hartree–Fock modes). The list below enumerates the most widely known of these codes:

- Non-relativistic codes with relativistic corrections:
    - C. Froese Fischer et al.: ATSP2K (MCHF) [14]
    - A. Hibbert: CIV3 [15]
    - W. Eissner et al.: SUPERSTRUCTURE [36]
    - P. Bogdanovich et al.: Quasirelativistic Hartree–Fock (CI; unpublished; see, e.g., [37])
- Relativistic codes:
    - I. Grant, C. Froese Fischer, P. Jönsson et al.: GRASP2K (MCDHF) [16,38]
    - M. F. Gu: FAC (MCDF, MBPT) [18]
    - J.P. Desclaux, P. Indelicato: MCDFGME [17]
    - Y. Ishikawa: Møller–Plesset many-body perturbation theory (MR-MP; unpublished; see, e.g., [19])
    - W.R. Johnson, U. I. Safronova, M.S. Safronova: RMBPT (unpublished, see, e.g., [39])
    - M.S. Safronova et al.: RMBPT (all-order; unpublished); see, e.g., [40]
    - E. Eliav, U. Kaldor: Coupled-Cluster (unpublished, see, e.g., [41])
    - V.A. Dzuba, W.R. Johnson: SD Coupled-Cluster (unpublished, see, e.g., [42])
    - I.M. Savukov: parametric CI+MBPT (unpublished, see, e.g., [43])

Some of the codes listed above possess limited options for semiempirical parametric fitting, for example, the CIV3 code [15], which allows for a semiempirical adjustment of the diagonal elements of the Hamiltonian matrix, roughly equivalent to the adjustment of the configuration average energies in Cowan's LSF. The SUPERSTRUCTURE code [36] allows for the adjustment of the $\lambda$ parameters of the radial wavefunctions. Savukov's CI+MBPT (configuration interaction plus many-body perturbation theory) method [43] also includes some free parameters that can be optimized by a least-squares fitting of experimental energy levels. However, I am not aware of any publicly available implementation of these semiempirical fitting procedures.

## 10. Known Problems in Cowan's Codes

One of the recently found problems in Cowan's codes is related to CIs between configurations forming a Rydberg series [44]. Brillouin's theorem [1] requires the direct CI integrals of rank zero of such configurations to be exactly zero. The problem arises because of the method used in Cowan's codes to identify a Rydberg series. In Cowan's implementation, it was thought that the Rydberg electron is always in the last occupied shell specified in the input file for RCN. However, it is not always possible to arrange the input file in this way. Moreover, most users do not even know about this rule and specify electronic shells using a standard ordering of the $nl$ shells, with the principal ($n$) and orbital ($l$) quantum numbers increasing from left to right. The following example gives two possible layouts of the RCN input file for calculating the odd-parity configuration complex ($1s^22p$ + $1s^23p$ + $1s^24p$ + $3p5p^2$ + $4p5p^2$) in neutral lithium:

(a)

```
3        1Li I s22p        1s2 5p0 2p
3        1Li I s23p        1s2 5p0 3p
3        1Li I s24p        1s2 5p0 4p
3        1Li I 5p23p       1s0 5p2 3p
3        1Li I 5p24p       1s0 5p2 4p
```

(b)

```
3        1Li I s22p        1s2 2p
3        1Li I s23p        1s2 2p0 3p
3        1Li I s24p        1s2 2p0 3p0 4p
3        1Li I 5p23p       1s0 2p0 3p  4p0 5p2
3        1Li I 5p24p       1s0 2p0 3p0 4p  5p2
```

Both layouts are allowed by RCN and produce the same values for all single-configuration Slater parameters. However, the CI parameters are (incorrectly) calculated to be different for the last two configurations. Since these configurations differ by only the principal quantum number of one electron (3p or 4p), they are Rydberg configurations by definition. Layout (a) correctly produces the CI integral $R_d{}^0(5p3p,5p4p) = 0$. However, layout (b) results in the corresponding CI integral $R_d{}^0(3p5p,4p5p) \neq 0$, which is incorrect. The "bug fix" suggested in the erratum of reference [44] solves the problem only partially, where it provides the solution to the problem occurring in most commonly encountered cases, but not in such as given in the above example.

Other known problems are related to the transformation between different coupling schemes:

- No transformation to *jj* and *jK* coupling is available for shells with equivalent electrons. Such shells are always represented in *LS* coupling.
- Recoupling to a non-standard order of shell summation is not exact in the presence of CI.

A workaround exists for both these problems, where one may use the 'Coupling' code designed by G. Gaigalas as part of a future version of the GRASP2K package [45]. This code can recouple any set of eigenvectors, including those produced by Cowan's code, to virtually any possible coupling scheme, including variations in the order of summation of shells.

## 11. Summary and Outlook

The failure of modern ab initio atomic theory methods to reproduce experimentally observed energy structures of complex heavy atoms and ions indicates that semiempirical analysis with parametric fitting (LSF) will remain the main tool for the theoretical interpretation of complex spectra, such as lanthanides and actinides. Cowan's codes remain the most universal tool in such analyses. However, these codes have many limitations. To overcome them, new LSF codes are needed that can better account for the relativistic and second-order effects, such as spin-spin and spin-orbit interactions. In particular, the orthogonal operators method needs to be extended to spectra with open f shells. As of today, no fully relativistic atomic codes exist that are capable of comprehensive parametric fitting. The development of such codes would be most welcome. Finally, codes for the parametric fitting of hyperfine structures and isotope shifts, some of which exist for private use, should be published and made available. These developments are needed to address the long-standing deficiency in experimental analyses of atomic spectra illustrated in the triangular diagram in Figure 6.

As seen in Figure 6, no experimental data exist for most spectra of atoms heavier than molybdenum with core charges $Z_c > 5$. Many of these unknown spectra are becoming of urgent interest for new applications, such as for atomic clocks on heavy multicharged ions [46], extreme ultraviolet lithography for the production of next-generation semiconductor circuits [47], and inertial fusion [48]. The development of accurate semiempirical atomic codes is of vital importance for the analysis of such spectra.

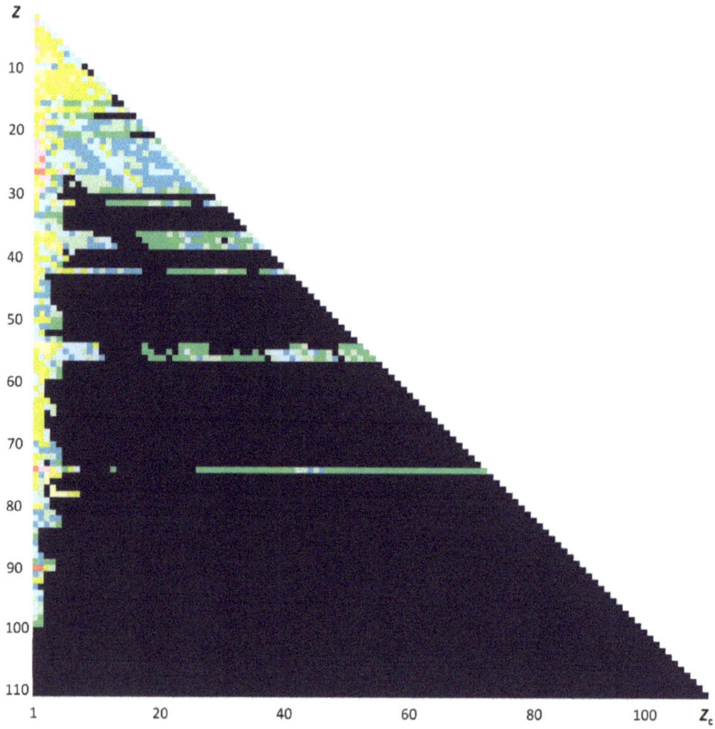


**Figure 6.** The Lines Holdings from the NIST Atomic Spectra Database [20]. Each square represents an ion with atomic number $Z$ and core charge $Z_c$ (e.g., $Z_c = 1$ represents neutral atoms, $Z_c = 2$ represents singly ionized, etc.). The color of the squares represents the amount of experimental data on the spectrum. The brighter the color, the more data are available. A black color represents that no data are available for a given spectrum.

**Funding:** This research received no external funding.

**Conflicts of Interest:** The author declares no conflict of interest.

## References

1. Cowan, R.D. *The Theory of Atomic Structure and Spectra*; University of California Press: Berkeley, CA, USA, 1981.
2. Cowan, R.D. Theoretical Calculation of Atomic Spectra Using Digital Computers. *J. Opt. Soc. Am.* **1968**, *58*, 808–818. [CrossRef]
3. Cowan, R.D. Atomic Self-Consistent-Field Calculations Using Statistical Approximations for Exchange and Correlation. *Phys. Rev.* **1967**, *163*, 54–61. [CrossRef]
4. Cowan, R.D.; Andrew, K.L. Coupling Considerations in Two-Electron Spectra. *J. Opt. Soc. Am.* **1965**, *55*, 502–516. [CrossRef]
5. Clark, C.A. Obituary: Robert Duane Cowan 26 July 2018. Available online: https://ladailypost.com/content/obituary-robert-duane-cowan-july-26-2018 (accessed on 9 October 2018).
6. Web of Science, Originally Produced by the Institute for Scientific Information (ISI), Now Maintained by Clarivate Analytics. Available online: http://isiknowledge.com/ (accessed on 10 September 2018).

7. Bordarier, Y.; Vetter, R.; Blaise, J. Étude des structures hyperfines des raies d'arc de $^{169}$Tm. *J. Phys.* **1963**, *24*, 1107–1112. [CrossRef]
8. Bordarier, Y.; Judd, B.R.; Klapisch, M. Hyperfine Structure of Eu I. *Proc. R. Soc. Lond. Ser. A* **1965**, *289*, 81–96.
9. Gluck, G.G.; Bordarier, Y.; Bauche, J.; van Kleef, T.A.M. Deplacement isotopique, calculs theoriques et structure des termes dans le spectre d'arc de l'osmium. *Physica* **1964**, *30*, 2068–2104. [CrossRef]
10. Carlier, A.; Blaise, J.; Schweighofer, M.-G. Étude des configurations impaires $4f^6 6s6p$ et $4f^5 5d6s^2$ de Sm I. *J. Phys.* **1968**, *29*, 729–738. [CrossRef]
11. Hansen, J.E.; Uylings, P.H.M.; Raassen, A.J.J. Parametric Fitting with Orthogonal Operators. *Phys. Scr.* **1988**, *37*, 664–672. [CrossRef]
12. Wyart, J.-F. On the Interpretation of Complex Atomic Spectra by Means of the Parametric Racah-Slater Method and Cowan Codes. *Can. J. Phys.* **2011**, *89*, 451–456. [CrossRef]
13. Kramida, A. Critical Evaluation of Data on Atomic Energy Levels, Wavelengths, and Transition Probabilities. *Fusion Sci. Technol.* **2013**, *63*, 313–323. [CrossRef]
14. Froese Fischer, C.; Tachiev, G.; Gaigalas, G.; Godefroid, M.R. An MCHF Atomic-Structure Package for Large-Scale Calculations. *Comput. Phys. Commun.* **2007**, *176*, 559–579. [CrossRef]
15. Hibbert, A. CIV3—A General Program to Calculate Configuration Interaction Wave Functions and Electric-Dipole Oscillator Strengths. *Comput. Phys. Commun.* **1975**, *9*, 141–172. [CrossRef]
16. Jönsson, P.; Gaigalas, G.; Bieroń, J.; Froese Fischer, C.; Grant, I.P. New Version: GRASP2K Relativistic Atomic Structure Package. *Comput. Phys. Commun.* **2013**, *184*, 2197–2203. [CrossRef]
17. Desclaux, J.P.; Indelicato, P. *Multiconfiguration Dirac-Fock with General Matrix Elements*; Laboratoire Kastler Brossel: Paris, France, 2017; Available online: http://dirac.spectro.jussieu.fr/mcdf/mcdf_code/mcdfgme_accueil.html (accessed on 1 July 2019).
18. Gu, M.F. The Flexible Atomic Code. *Can. J. Phys.* **2008**, *86*, 675–689.
19. Vilkas, M.J.; Koc, K.; Ishikawa, Y. Relativistic Multireference Møller-Plesset Perturbation Theory Based on Multiconfigurational Dirac-Fock Reference Functions. *Chem. Phys. Lett.* **1998**, *296*, 68–76. [CrossRef]
20. Kramida, A.; Ralchenko, Y.; Reader, J.; NIST ASD Team. *NIST Atomic Spectra Database (Ver. 5.6.1)*; National Institute of Standards and Technology: Gaithersburg, MD, USA, 2018. Available online: https://physics.nist.gov/asd (accessed on 1 July 2019).
21. McGuinness, C.; Robert, D. *Cowan's Atomic Structure Code*; Trinity College, University of Dublin: Dublin, Ireland, 2007; Available online: https://www.tcd.ie/Physics/people/Cormac.McGuinness/Cowan/ (accessed on 1 July 2019).
22. Kramida, A. *A Suite of Atomic Structure Codes Originally Developed by RD Cowan Adapted for Windows-Based Personal Computers*; National Institute of Standards and Technology: Gaithersburg, MD, USA, 2018. [CrossRef]
23. Quinet, P.; Palmeri, P.; Biémont, E. On the use of the Cowan's code for atomic structure calculations in singly ionized lanthanides. *J. Quant. Spectrosc. Radiat. Transf.* **1999**, *62*, 625–646. [CrossRef]
24. Kurucz, R.L. *Atoms*; Harvard University: Cambridge, MA, USA, 2017; Available online: http://kurucz.harvard.edu/atoms.html (accessed on 1 July 2019).
25. Ruczkowski, J.; Elantkowska, M.; Dembczyński, J. An Alternative Method for Determination of Oscillator Strengths: The Example of Sc II. *J. Quant. Spectrosc. Radiat. Transf.* **2014**, *145*, 20–42. [CrossRef]
26. Abdallah, J., Jr.; Clark, R.E.H.; Cowan, R.D. *Theoretical Atomic Physics Code Development I. CATS: Cowan Atomic Structure Code*; Report LA-11436-M; Los Alamos National Lab: Los Alamos, NM, USA, 1988; Volume I, p. 31. Available online: https://fas.org/sgp/othergov/doe/lanl/docs3/00323386.pdf (accessed on 1 July 2019).
27. Azarov, V.I.; Kramida, A.; Vokhmentsev, M.Y. IDEN2–A Program for Visual Identification of Spectral Lines and Energy Levels in Optical Spectra of Atoms and Simple Molecules. *Comput. Phys. Commun.* **2018**, *225*, 149–153. [CrossRef]
28. Raassen, A.J.J.; Uylings, P.H.M. The Use of Complete Sets of Orthogonal Operators in Spectroscopic Studies. *Phys. Scr.* **1996**, *65*, 84–87. [CrossRef]
29. van het Hof, G.J.; Raassen, A.J.J.; Uylings, P.H.M. Parametric Description of $3d^N 4s$ Configurations using Orthogonal Operators. *Phys. Scr.* **1991**, *44*, 343–350.
30. Azarov, V.I.; Gayasov, R.R. The Third Spectrum of Rhenium (Re III): Analysis of the $(5d^5+5d^4 6s)$–$(5d^4 6p+5d^3 6s6p)$ Transition Array. *At. Data Nucl. Data Tables* **2018**, *122*, 306–344. [CrossRef]

31. Azarov, V.I.; Tchang-Brillet, W.-Ü.L.; Gayasov, R.R. Analysis of the Spectrum of the ($5d^6$+$5d^56s$)–($5d^56p$+$5d^46s6p$) Transitions of Two Times Ionized Osmium (Os III). *At. Data Nucl. Data Tables* **2018**, *122*, 345–377. [CrossRef]
32. Wyart, J.-F.; Raassen, A.J.J.; Uylings, P.H.M.; Joshi, Y.N. Spectra of High-*Z* Ions of Stellar Interest. A Theoretical Study of $(d + s)^8$ Mixed Configurations in 5d- and 4d-Elements. *Phys. Scr.* **1993**, *47*, 59–64. [CrossRef]
33. Uylings, P.H.M.; (University of Amsterdam). Private communication, 2018.
34. Blaise, J.; Wyart, J.-F.; Conway, J.G.; Worden, E.F. Generalized Parametric Study of $5f^N$ and $5f^N7s$ Configurations. *Phys. Scr.* **1980**, *22*, 224–230. [CrossRef]
35. Wyart, J.-F.; Raassen, A.J.J.; Joshi, Y.N.; Uylings, P.H.M. The $5d^96p$-$5d^9(6d+7s)$ Transitions in the Isoelectronic Sequence Au II-Bi VI. *J. Phys. II* **1992**, *2*, 895–912. [CrossRef]
36. Eissner, W.; Jones, M.; Nussbaumer, H. Techniques for the Calculation of Atomic Structures and Radiative Data Including Relativistic Corrections. *Comput. Phys. Commun.* **1974**, *8*, 270–306. [CrossRef]
37. Bogdanovich, P.; Rancova, O. Quasirelativistic Hartree-Fock Equations Consistent with Breit-Pauli Approach. *Phys. Rev.* **2006**, *74*, 052501. [CrossRef]
38. Froese Fischer, C.; Gaigalas, G.; Jönsson, P.; Bieroń, J. GRASP2018—A Fortran 95 version of the General Relativistic Atomic Structure Package. *Comput. Phys. Commun.* **2019**, *237*, 184–187. [CrossRef]
39. Safronova, M.S.; Johnson, W.R.; Safronova, U.I. Relativistic Many-Body Calculations of the Energies of $n = 2$ States for the Berylliumlike Isoelectronic Sequence. *Phys. Rev.* **1996**, *53*, 4036–4053. [CrossRef]
40. Safronova, U.I.; Safronova, M.S.; Kozlov, M.G. Relativistic All-Order Calculations of In I and Sn II Atomic Properties. *Phys. Rev.* **2007**, *76*, 022501. [CrossRef]
41. Eliav, E.; Kaldor, U.; Ishikawa, Y. Ionization Potentials and Excitation Energies of the Alkali-Metal Atoms by the Relativistic Coupled-Cluster Method. *Phys. Rev. A* **1994**, *50*, 1121–1128. [CrossRef] [PubMed]
42. Dzuba, V.A.; Johnson, W.R. Coupled-Cluster Single-Double Calculations of the Relativistic Energy Shifts in C IV, Na I, Mg II, Al III, Si IV, Ca II, and Zn II. *Phys. Rev. A* **2007**, *76*, 062510. [CrossRef]
43. Savukov, I.M. Parametric CI+MBPT Calculations of Th I Energies and *g*-Factors for Even States. *J. Phys. B* **2007**, *50*, 165001. [CrossRef]
44. Kramida, A. Configuration Interactions of Class 11: An Error in Cowan's Atomic Structure Theory. *Comput. Phys. Commun.* **2017**, *215*, 47–48, *Erratum* **2018**, *232*, 266–267. [CrossRef]
45. Gaigalas, G. Coupling: The program for searching optimal coupling scheme in atomic theory. *Submitted to Comput. Phys. Commun.* **2019**.
46. Safronova, M.S.; Dzuba, V.A.; Flambaum, V.V.; Safronova, U.I.; Porsev, S.G.; Kozlov, M.G. Highly Charged Ions for Atomic Clocks, Quantum Information, and Search for $\alpha$ variation. *Phys. Rev. Lett.* **2014**, *113*, 030801. [CrossRef]
47. Banine, V.Y.; Koshelev, K.N.; Swinkels, G.H.P.M. Physical Processes in EUV Sources for Microlithography. *J. Phys. D* **2011**, *44*, 253001. [CrossRef]
48. Glenzer, S.H.; Fournier, K.B.; Wilson, B.G.; Lee, R.W.; Suter, L.J. Ionization Balance in Inertial Confinement Fusion Hohlraums. *Phys. Rev. Lett.* **2001**, *87*, 045002. [CrossRef] [PubMed]

atoms

Article

# Electron-Impact Dissociation of Vibrationally-Excited Molecular Hydrogen into Neutral Fragments

**Liam H. Scarlett [1,*], Jeremy S. Savage [1], Dmitry V. Fursa [1], Mark C. Zammit [2] and Igor Bray [1]**

1 Curtin Institute for Computation and Department of Physics and Astronomy, Curtin University, Perth, WA 6102, Australia
2 Theoretical Division, Los Alamos National Laboratory, Los Alamos, NM 87545, USA
* Correspondence: liam.scarlett@postgrad.curtin.edu.au

Received: 29 June 2019; Accepted: 2 August 2019; Published: 6 August 2019

**Abstract:** We present convergent close-coupling (CCC) calculations of electron-impact dissociation of vibrationally-excited molecular hydrogen into neutral fragments. This work follows from our previous results for dissociation of molecular hydrogen in the ground vibrational level [Scarlett et al., Eur. Phys. J. D **72**, 34 (2018)], which were obtained from calculations performed in a spherical coordinate system. The present calculations, performed utilizing a spheroidal formulation of the molecular CCC method, reproduce the previous dissociation cross sections for the ground vibrational level, while allowing the extension to scattering on excited levels.

**Keywords:** hydrogen molecule; electron-impact excitation; dissociation

## 1. Introduction

The dissociation of molecular hydrogen by electron-impact excitation is a process of significant importance in the modeling of hydrogenic plasmas. The $H_2$ molecule is abundant in astrophysical gas clouds and plasmas, where molecular dissociation has implications for stellar formation and atmospheric modeling [1]. The neutral hydrogen found in the divertor region of the tokamak fusion reactors is primarily molecular. The rate at which it dissociates plays a role in determining the properties of the edge plasma, which in turn affects the performance of the core fusion plasma [2,3].

Many reaction channels lead to dissociation, which can be grouped into the following two general processes:

$$\begin{aligned} e^- + H_2 \rightarrow H_2^* + e^- &\rightarrow H(n\ell) + H(n'\ell') + e^- && \text{(1 - dissociation into neutral fragments)} \\ &\rightarrow H_2^{+*} + 2e^- \rightarrow H^+ + H(n\ell) + 2e^-. && \text{(2 - dissociative ionization)} \end{aligned}$$

In this paper, we focus our attention on the first process—dissociation leading to only neutral atomic fragments. This process is challenging to study experimentally, since the detection of neutral fragments is substantially more difficult than it is for charged fragments. The only available experimental data [4] for dissociation of the ground vibrational level were obtained by subtracting the ionization cross section from Corrigan's 1965 measurements [5] of the total dissociation cross section (including Processes (1) and (2)). These measurements were performed more than 50 years ago and there have been no subsequent attempts at repeating them. For dissociation of vibrationally-excited $H_2$, there are no experimental data.

Theoretical estimates of dissociation cross sections are difficult due to the large number of dissociation channels which must be accounted for. In a previous paper [6], we analyzed the convergent close-coupling (CCC) results for electronic excitation of $H_2$ [7] to produce a cross section for dissociation of $H_2$ in the ground electronic and vibrational state, finding that several hundred electronic states must be accounted for to yield a convergent dissociation cross section. Furthermore, a proper treatment of dissociation (including excitation-radiative-decay dissociation) requires collision data resolved not only in the electronic levels, but in the initial and final vibrational levels, which vastly increases the computational expense of the calculations. Prior to the CCC estimates of dissociation [6], calculations were limited to low incident energies where only a small number of reaction channels are open [8–10], or large energies where the Born approximation may be applied [11]. No attempts were made to produce a dissociation cross section including all reaction channels over a wide range of incident energies. The low-energy adiabatic-nuclei (AN) $R$-matrix calculations of the $X^1\Sigma_g^+ \to b^3\Sigma_u^+$ excitation by Stibbe and Tennyson [10] provide a good estimate of the dissociation cross section below approximately 12 eV (depending on the initial vibrational level), where the higher electronic states are closed. These are the only previous quantum-mechanical calculations which account for scattering on excited vibrational levels. However, due to the difficulty of performing molecular scattering calculations at large internuclear separations, the $R$-matrix calculations only treated scattering on the $v_i = 0$–$4$ vibrational levels explicitly, using an extrapolation procedure for the remaining vibrational levels [12].

Recently, the molecular CCC method has been utilized to produce cross sections for dissociative excitation [13] and excitation-radiative-decay dissociation [14], for scattering on all $v_i = 0$–$14$ bound vibrational levels of the ground electronic ($X^1\Sigma_g^+$) state to a number of low-lying singlet states of $H_2$. To allow accurate structure and scattering calculations to be performed over the range of internuclear separations spanned by the $v_i = 0$–$14$ vibrational levels, the CCC theory was formulated in spheroidal coordinates [15]. Here, we utilize these results, as well as calculations for excitation of vibrationally-excited $H_2$ into the triplet system, to produce $e^-$–$H_2$ dissociation cross sections for scattering on all initial vibrational levels, over the energy range from threshold to 120 eV. Atomic units are used throughout the paper unless specified otherwise.

## 2. Theory

Excitation-induced molecular dissociation occurs via three main mechanisms: dissociative excitation (DE), excitation-radiative-decay dissociation (ERDD), and predissociation (PD). Detailed discussions of the spheroidal molecular CCC method, and the calculations of DE and ERDD cross sections can be found in Refs. [6,13–15], thus only a brief overview is given here.

### 2.1. Spheroidal Molecular CCC Method

The spheroidal molecular CCC method follows the same approach as the spherical-coordinate implementation, for which a detailed discussion can be found in Ref. [16]. The principle difference is that the coordinate space of the projectile-target system is described using a system of prolate spheroidal coordinates $\boldsymbol{\rho} = (\rho, \eta, \phi)$, with

$$\rho = \frac{r_1 + r_2}{2} - \frac{R}{2} \text{ and } \eta = \frac{r_1 - r_2}{2}, \tag{1}$$

where $R$ is the internuclear separation, and $r_1$ and $r_2$ are the distances from the two nuclei. The $z$-axis is aligned with the internuclear axis, and the azimuthal angle $\phi$ retains its definition from the spherical coordinate system. The use of spheroidal coordinates provides a significant improvement in both accuracy and efficiency when calculating the target wave functions, which are inherently non-spherical, particularly at larger values of $R$.

The Born–Oppenheimer approximation is utilized, allowing the electronic structure calculation to be performed independent of the nuclear motion, at fixed values of $R$. The target wave functions and energies are obtained by diagonalizing the fixed-nuclei electronic molecular Hamiltonian in a basis of Sturmian (Laguerre) functions, which is optimized to yield an accurate description of the low-lying states, and an adequate pseudo-state discretization of the continuum. Specific details of the structure calculation used in the present work can be found in Ref. [15].

The projectile wave functions are expanded in partial waves of spheroidal pseudo-angular-momentum $\lambda$ and angular-momentum projection $m$, and the total scattering wave function is expanded in terms of the asymptotic channels of the scattering system Hamiltonian. The partial-wave fixed-nuclei scattering amplitudes $F_{f\lambda_f m_f, i\lambda_i m_i}(R; E_{\text{in}})$ are obtained by solving the resulting set of coupled Lippmann–Schwinger equations at a given incident electron energy $E_{\text{in}}$ and internuclear separation $R$. The number of target states and projectile partial waves included in the calculations determines the accuracy of the scattering cross sections. The present calculations utilize a scattering model consisting of 210 bound and continuum electronic (pseudo)states, which we have found to yield a sufficient level of convergence for a number of low-lying excitations. A full discussion of the scattering models we have used and the methods for obtaining the scattering amplitudes can be found in Refs. [13,15].

The adiabatic-nuclei approximation is invoked to restore the dependence on the nuclear motion, yielding cross sections for vibrationally-resolved electronic transitions $iv_i \to fv_f$:

$$\sigma_{fv_f, iv_i}(E_{\text{in}}) = \frac{q_{fv_f}}{4\pi q_i} \sum_{\substack{\lambda_f, \lambda_i \\ m_f, m_i}} \left| \langle \nu_{fv_f} | F_{f\lambda_f m_f, i\lambda_i m_i}(R; E_{\text{in}}) | \nu_{iv_i} \rangle \right|^2, \tag{2}$$

where $q_{fv_f}$ and $q_i$ are the outgoing and incident projectile linear momenta, and $\nu_{nv_n}$ are the vibrational wave functions, which are obtained by diagonalizing the Born–Oppenheimer nuclear Hamiltonian

$$H_n^{\text{BO}} = -\frac{1}{2\mu} \frac{\text{d}^2}{\text{d}R^2} + \frac{J(J+1) - \Lambda_n^2}{2\mu R^2} + \epsilon_n(R), \tag{3}$$

where $\mu$ is the nuclear reduced mass, $\epsilon_n$ is the potential-energy curve of the electronic state $n$, $J$ is the total molecular angular momentum, and $\Lambda_n$ is the electronic-state angular momentum projection onto the internuclear axis. For the purposes of obtaining cross sections resolved in the target electronic and vibrational levels only, we may neglect the centrifugal term in Equation (3), as for small $J$ it is negligible compared to $\epsilon_n$.

It is worth noting that the AN cross section we define in Equation (2) includes the correct outgoing projectile momentum

$$q_{fv_f} = \sqrt{2\left[E_{\text{in}} - \varepsilon_{fv_f, iv_i}\right]}, \tag{4}$$

where $\varepsilon_{fv_f, iv_i}$ is the vibrationally-resolved excitation energy. The more common AN formulation (as given by Lane [17], for example) absorbs the outgoing momentum into the integral over $R$, in doing so replacing it with the $R$-dependent fixed-nuclei momentum

$$q_f(R) = \sqrt{2\left[E_{\text{in}} - \epsilon_{f,i}(R)\right]}, \tag{5}$$

where $\epsilon_{f,i}$ is the fixed-nuclei (vertical) excitation energy. In the present notation, the resulting AN cross section is

$$\sigma_{fv_f, iv_i}(E_{\text{in}}) = \frac{1}{4\pi q_i} \sum_{\substack{\lambda_f, \lambda_i \\ m_f, m_i}} \left| \langle \nu_{fv_f} | \sqrt{q_f(R)} F_{f\lambda_f m_f, i\lambda_i m_i}(R; E_{\text{in}}) | \nu_{iv_i} \rangle \right|^2. \tag{6}$$

The latter method allows for the electronic excitation cross section (summed over final vibrational levels) to be evaluated using

$$\sigma_{f, iv_i} = \langle \nu_{iv_i} | \sigma_{f,i} | \nu_{iv_i} \rangle, \tag{7}$$

where $\sigma_{f,i}$ is the ($R$-dependent) FN electronic excitation cross section. In the present formulation, Equation (2) must be summed over final levels numerically, however at larger incident energies ($\approx$10 eV above the electronic excitation threshold), Equation (7) returns the same result. The present formulation in Equation (2) is convenient when partitioning excitation cross sections into the various dissociative pathways as it retains the correct vibrationally-resolved excitation thresholds, which are lost in the standard AN formulation.

### 2.2. Dissociative Excitation

DE is the direct impact-induced transition from the initial state $iv_i$ to a dissociative level in the vibrational continuum of some electronic state $f$. Evaluation of the DE cross section formally requires integration of the kinetic-energy-release cross section $\mathrm{d}\sigma_{f,iv_i}/\,\mathrm{d}E_\mathrm{k}$ over the kinetic energy $E_\mathrm{k}$ of the dissociation fragments. However, the dissociative pseudostates resulting from diagonalizing Equation (3) form a quadrature rule for integrating over the vibrational continuum, so we can equivalently sum Equation (2) over these states in the same way that ionization is typically treated in the CCC method. We have previously performed calculations for dissociative excitation of the $B^1\Sigma_u^+$, $C^1\Pi_u$, $B'^1\Sigma_u^+$, $D^1\Pi_u$, and $E,F^1\Sigma_g^+$ electronic singlet states for electrons scattering on all $v_i$ = 0–14 bound vibrational levels of the $H_2$ ground electronic state ($X^1\Sigma_g^+$) [13].

### 2.3. Excitation-Radiative-Decay Dissociation

ERDD proceeds via excitation of the bound vibrational spectrum of an excited electronic state, followed by radiative decay to the dissociative vibrational continuum of a lower electronic state. The decay sequences can include multiple electronic states, and terminate in the bound spectrum of the $X^1\Sigma_g^+$ state or the vibrational continuum of either the ground or an excited electronic state. We have adopted the following approach [14] for obtaining the ERDD cross section for an electronic state $f$: we calculate the excitation-radiative-decay cross section for decays back to the bound vibrational levels of the ground electronic state

$$\sigma^{\mathrm{ERD}}_{iv_i',f,iv_i} = (1 - F^{\mathrm{PD}}_{f,iv_i}) \sum_{v_f} \frac{A_{i,f}(v_i', v_f)}{A_f(v_f)} \sigma_{fv_f,iv_i} \tag{8}$$

and subtract it from the cross section for excitation of the bound levels of state $f$, giving

$$\sigma^{\mathrm{ERDD}}_{i,f,iv_i} = (1 - F^{\mathrm{PD}}_{f,iv_i}) \sum_{v_f} \left( 1 - \sum_{v_i'} \frac{A_{i,f}(v_i', v_f)}{A_f(v_f)} \right) \sigma_{fv_f,iv_i}. \tag{9}$$

In Equations (8) and (9), $A_{i,f}(v_i', v_f)$ is the $fv_f \to iv_i'$ radiative transition probability, $A_f(v_f)$ is the total transition probability for the state $fv_f$, and $F^{\mathrm{PD}}_{f,iv_i}$ is the fraction of the excitation cross section $\sigma_{f,iv_i}$ which leads to PD. We have previously performed calculations of excitation-radiative-decay dissociation for the $B^1\Sigma_u^+$, $C^1\Pi_u$, $B'^1\Sigma_u^+$, $D^1\Pi_u$, and $E,F^1\Sigma_g^+$ electronic singlet states, for electrons scattering on all $v_i$ = 0–14 bound vibrational levels of the $X^1\Sigma_g^+$ state [14].

### 2.4. Constructing the Dissociation Cross Section

The contribution from the electronic singlet spectrum to the total neutral-fragment dissociation cross section is the sum of the DE, ERDD, and PD processes for all electronically bound (i.e., non-ionizing) singlet states. For the $B^1\Sigma_u^+$, $C^1\Pi_u$, $B'^1\Sigma_u^+$, $D^1\Pi_u$, and $E,F^1\Sigma_g^+$ states we use the CCC DE and ERDD cross sections which have been previously calculated [13,14]. PD does not occur or is negligible for these states, except for the $D^1\Pi_u$ state. We take the PD fraction of 0.2564 for the $D^1\Pi_u$ state from Glass-Maujean et al. [18]. To approximately account for dissociation through the remaining singlet states, we obtain an effective dissociation fraction for the states we have considered explicitly by summing their respective dissociation cross sections and dividing by their summed excitation cross section. We then apply this dissociation fraction to each of the remaining excited bound electronic states. The same approach to approximately accounting for higher excited singlet states was used in our previous calculations of dissociation from the ground vibrational level [6]. Although dissociation may also proceed through direct excitation of the $X^1\Sigma_g^+$ vibrational continuum, we have found that the cross sections for this process are two orders of magnitude smaller than the $b^3\Sigma_u^+$ DE cross section for scattering on the same initial vibrational level, and are therefore neglected.

Dissociation through the triplet spectrum is simpler to analyse. Assuming an *L*-*S* coupling scheme, radiative transitions between the triplet spectrum and the singlet ground state are spin forbidden. Since the lowest triplet state of $H_2$ ($b^3\Sigma_u^+$) is purely repulsive, excitation of any triplet state must eventually result in dissociation. The cross section for dissociation through the triplet spectrum is then simply the sum of all electronically bound triplet-state excitations.

Stibbe and Tennyson [10] proposed an *energy-balancing* modification to the AN method, which more accurately treats the partitioning of energies between the outgoing electron and dissociation fragments. The technique requires evaluating the fixed-nuclei scattering amplitudes at a modified *R*-dependent incident energy

$$E(R) = E_{\text{in}} - \varepsilon_{fv_f,iv_i} + \epsilon_{f,i}(R). \tag{10}$$

This definition ensures that at every value of $R$ the fixed-nuclei outgoing momentum $q_f(R)$ is equal to the physical outgoing momentum $q_{fv_f}$ (Equations (4) and (5) are equivalent when $E_{\text{in}}$ is replaced by $E(R)$ in Equation (5)). In the present notation, the energy-balanced cross section is

$$\sigma^{\text{e bal}}_{fv_f,iv_i}(E_{\text{in}}) = \frac{q_{fv_f}}{4\pi q_i} \sum_{\substack{\lambda_f,\lambda_i \\ m_f,m_i}} \left| \langle \nu_{fv_f} | F_{f\lambda_f m_f,i\lambda_i m_i}(R; E(R)) | \nu_{iv_i} \rangle \right|^2. \tag{11}$$

Stibbe and Tennyson [10] applied this technique to the dissociative $b^3\Sigma_u^+$ excitation using true continuum nuclear functions for the final vibrational levels, and integrating over the fragments kinetic energy to obtain integral DE cross sections. As the $b^3\Sigma_u^+$ excitation is the primary dissociation channel at low energies we utilize the energy-balancing technique for this transition, but rather than integrating the kinetic-energy-release cross section we equivalently sum Equation (11) over final vibrational pseudostates until convergence is reached. For the remaining triplet states, a less accurate approach is sufficient, and we utilize the analytical sum over final vibrational levels in Equation (7) to give

$$\sigma^{\text{rem. trip.}}_{iv_i} = \langle \nu_{iv_i} | \sum_f \sigma_{f,i} | \nu_{iv_i} \rangle, \tag{12}$$

where the sum is over all bound electronic triplet states above the $b^3\Sigma_u^+$ state.

## 3. Results

In Figure 1, we present the cross sections for excitation of the $b^3\Sigma_u^+$ state from all bound vibrational levels in the $X^1\Sigma_g^+$ state. This transition is the dominant dissociative process within the triplet system, and the dominant overall dissociative process at low energies. The cross section is substantially enhanced at low energies when scattering on excited vibrational levels. This enhancement is due to the degeneracy of the $X^1\Sigma_g^+$ and $b^3\Sigma_u^+$ states at larger internuclear separations, which results in small excitation energies for transitions from the high-lying vibrational levels of the ground state to the dissociative levels of the $b^3\Sigma_u^+$ state. As shown by Trevisan and Tennyson [19,20], a significant portion of the energy required to excite this transition is carried away by the dissociation fragments. The large cross section for this process (particularly for low-energy electrons scattering on excited vibrational levels) then corresponds to the release of significant quantities of hot H atoms into the plasma or other medium.

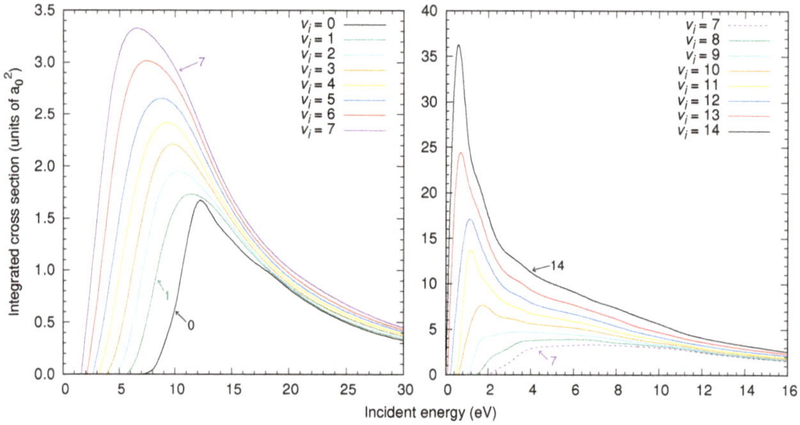

**Figure 1.** Electron-impact cross sections for excitation from the $v_i = 0$–14 vibrational levels of the $X^1\Sigma_g^+$ state to the dissociative $b^3\Sigma_u^+$ state of $H_2$. The cross sections increase monotonically with increasing $v_i$.

The summed cross section for excitation of the remaining triplet states is presented in Figure 2.

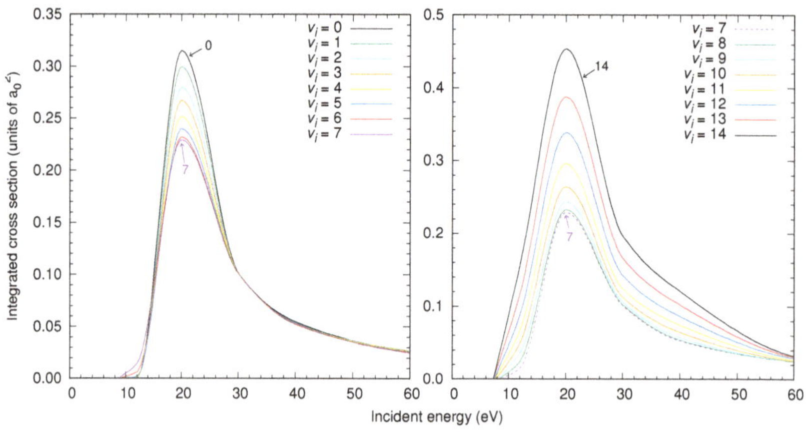

**Figure 2.** Electron-impact cross sections for excitation from the $v_i = 0$–14 vibrational levels of the $X^1\Sigma_g^+$ state to the bound electronic triplet spectrum of $H_2$ (minus the $b^3\Sigma_u^+$ state). The cross sections decrease with increasing $v_i$ up to $v_i = 7$ (**left**), and rise with increasing $v_i$ for $v_i > 7$ (**right**).

The remaining-triplet cross section is qualitatively similar to the $b^3\Sigma_u^+$ cross section for scattering on the ground vibrational level, but does not exhibit the low-energy spike for scattering on high vibrational levels, due to the electronic separation of the $X^1\Sigma_g^+$ state from the triplet spectrum above the $b^3\Sigma_u^+$ state. For the same reason, the cross sections for excitation of the remaining triplet states show a much smaller dependence on the initial vibrational level than the $b^3\Sigma_u^+$ excitation.

The contribution to the dissociation cross section from the singlet spectrum is presented in Figure 3. These results are the sum of the DE and ERDD cross sections published previously [13,14] for the $B^1\Sigma_u^+$, $C^1\Pi_u$, $B'^1\Sigma_u^+$, $D^1\Pi_u$, and $E,F^1\Sigma_g^+$ states, plus the estimate of PD through the $D^1\Pi_u$ state, and the estimate of dissociation through the remaining singlet states (as discussed in Section 2.4).

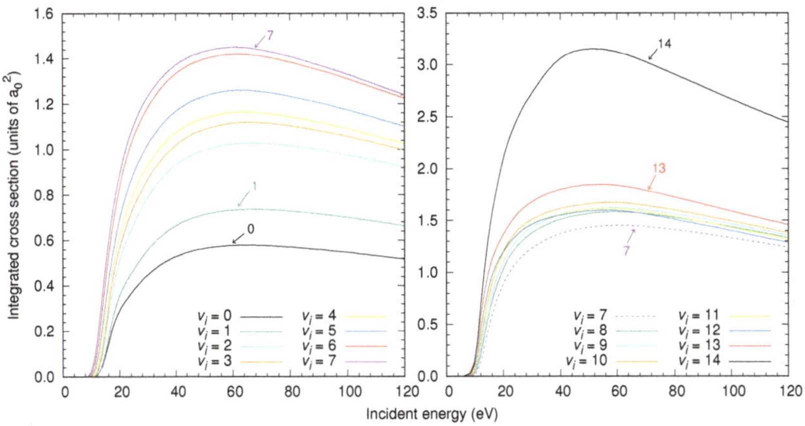


**Figure 3.** Electron-impact cross sections for dissociation of the $v_i = 0$–14 vibrational levels of the $X^1\Sigma_g^+$ state through the singlet spectrum of $H_2$.

The total neutral-fragment dissociation cross section is obtained by summing the above results for dissociation through the triplet and singlet spectra. For scattering on the ground vibrational level, we find agreement between the present results and our previous calculations [6] which were performed using converged (491-state) spherical-coordinate fixed-nuclei cross sections weighted with dissociation fractions obtained from a smaller (27-state) spheroidal adiabatic-nuclei model. In Figure 4, we compare the present (210-state) and previous (410-state) calculations with the recommended cross section of Yoon et al. [4]. The agreement between the two CCC results demonstrates that the present model yields sufficient convergence in the dissociation cross section. As we noted in our previous work [6], the substantial (up to a factor of 2) disagreement between the recommended data and the CCC results between 15 and 60 eV highlights the need for new measurements to be taken to clarify the situation.

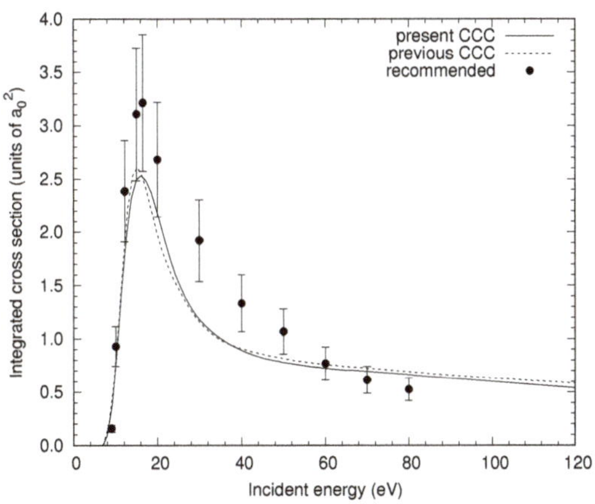


**Figure 4.** Electron-impact dissociation cross section for scattering on the ground (electronic and vibrational) level of $H_2$, comparing the present results with our previous calculations [6] and the recommended cross sections of Yoon et al. [4].

In Figure 5, we present the CCC results for dissociation of all $v_i = 0$–14 vibrational levels of the $X^1\Sigma_g^+$ state into neutral fragments. At low energies, the dissociation cross section is entirely comprised of the $b^3\Sigma_u^+$ excitation. The unusual shapes present in some curves are the result of higher electronic-state excitations becoming open as the incident energy increases. Over the entire energy range, the dissociation cross sections are significantly enhanced for scattering on excited vibrational levels, illustrating the importance of vibrationally-resolved collision data for modeling environments where molecules are present in vibrationally-excited states, such as fusion plasmas.

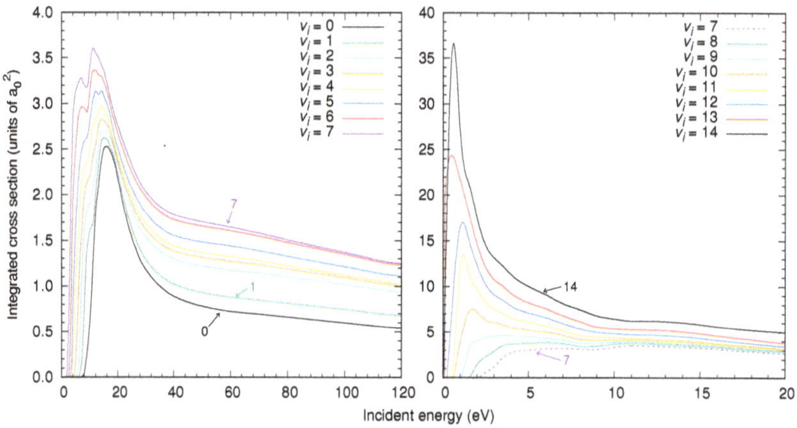


**Figure 5.** Electron-impact dissociation cross section for scattering on all bound vibrational levels of the $H_2$ ground electronic state. The cross sections rise monotonically with increasing vibrational level $v_i$.

The approximate way in which we have treated dissociation through the higher excited singlet states introduces an uncertainty in the dissociation results we have presented here. To provide the most liberal estimate, we allow for a 100% uncertainty in the contribution from the these singlets. We also estimate an overall uncertainty $u_{\mathrm{CCC}}$ of approximately 10% in the underlying CCC excitation cross sections, due to convergence and target structure accuracy. The combined uncertainty $u = \sqrt{u^2_{\mathrm{singlets}} + u^2_{\mathrm{CCC}}}$ is presented in Figure 6 as a function of incident energy and initial vibrational level. For scattering on the $v_i = 0$–7 vibrational levels, the present results have an uncertainty no greater than 15%, and up to $v_i = 12$ the uncertainty is less than 20%.

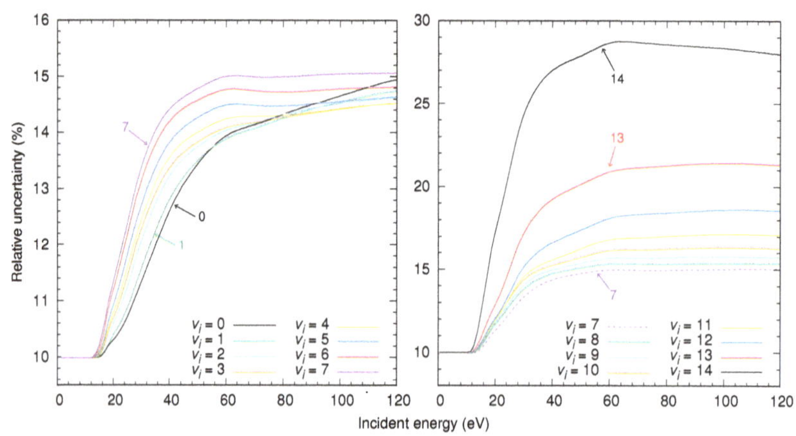


**Figure 6.** Relative uncertainty in the CCC dissociation cross section for each initial vibrational level.

## 4. Conclusions

We have presented a set of cross sections for electron-impact dissociation of vibrationally-excited $H_2$ into neutral fragments. An accurate treatment of the major dissociative mechanisms requires a fully vibrationally-resolved description of the $e^-$–$H_2$ scattering problem, which is made possible with the formulation of the molecular CCC method in spheroidal coordinates. For scattering on the ground vibrational level, the CCC results predict a substantially lower dissociation cross section than the only available experimental data in the 15–60 eV energy range [4,5], while for scattering on excited levels there are no available experimental data. We hope the present calculations, which represent the only available dissociation estimates for all initial vibrational levels over a wide range of incident energies, will be of interest for modeling fusion and astrophysical plasmas where molecular hydrogen is found in a range of vibrationally excited states.

**Author Contributions:** Methodology, L.H.S., J.S.S., M.C.Z., D.V.F. and I.B; writing–original draft preparation, L.H.S; writing–review and editing, L.H.S., J.S.S., M.C.Z., D.V.F. and I.B.

**Funding:** This work was supported by the United States Air Force Office of Scientific Research, Los Alamos National Laboratory (LANL), Curtin University, and resources provided by the Pawsey Supercomputing Centre, with funding from the Australian Government and Government of Western Australia. LHS acknowledges the contribution of an Australian Government Research Training Program Scholarship, and the support of the Forrest Research Foundation. MCZ would like to specifically acknowledge LANL's ASC PEM Atomic Physics Project for its support. LANL is operated by Triad National Security, LLC, for the National Nuclear Security Administration of the U.S. Department of Energy under Contract No. 89233218NCA000001.

**Conflicts of Interest:** The authors declare no conflict of interest.

## References

1. Liu, X.; Ahmed, S.M.; Multari, R.A.; James, G.K.; Ajello, J.M. High-resolution electron-impact study of the far-ultraviolet emission spectrum of molecular hydrogen. *Astrophys. J. Suppl. Ser.* **1995**, *101*, 375–399. [CrossRef]
2. Sawada, K.; Fujimoto, T. Effective ionization and dissociation hydrogen in plasma. *J. Appl. Phys.* **1995**, *78*, 2913–2924. [CrossRef]
3. Jonin, C.; Liu, X.; Ajello, J.M.; James, G.K.; Abgrall, H. High resolution electron-impact emission spectrum of $H_2$. I. Cross sections and emission yields 900–1200 Å. *Astrophys. J. Suppl. Ser.* **2000**, *129*, 247–266. [CrossRef]
4. Yoon, J.S.; Song, M.Y.; Han, J.M.; Hwang, S.H.; Chang, W.S.; Lee, B.; Itikawa, Y. Cross Sections for Electron Collisions with Hydrogen Molecules. *J. Phys. Chem. Ref. Data* **2008**, *37*, 913–931. [CrossRef]
5. Corrigan, S.J.B. Dissociation of Molecular Hydrogen by Electron Impact. *J. Chem. Phys.* **1965**, *43*, 4381. [CrossRef]
6. Scarlett, L.H.; Tapley, J.K.; Fursa, D.V.; Zammit, M.C.; Savage, J.S.; Bray, I. Electron-impact dissociation of molecular hydrogen into neutral fragments. *Eur. Phys. J. D* **2018**, *72*, 34. [CrossRef]
7. Zammit, M.C.; Savage, J.S.; Fursa, D.V.; Bray, I. Electron-impact excitation of molecular hydrogen. *Phys. Rev. A* **2017**, *95*, 022708. [CrossRef]
8. Chung, S.; Lin, C.C. Application of the close-coupling method to excitation of electronic states and dissociation of $H_2$ by electron impact. *Phys. Rev. A* **1978**, *17*, 1874–1891. [CrossRef]
9. Borges, I.; Jalbert, G.; Bielschowsky, C.E. Non-Franck-Condon electron-imact dissociative-excitation cross sections of molecular hydrogen producing $\mathrm{H}(1s) + \mathrm{H}(2l)$ through $X^1\Sigma_g^+(v=0) \rightarrow \{B^1\Sigma_u^+, B'^1\Sigma_u^+, C^1\Pi_u\}$. *Phys. Rev. A* **1998**, *57*, 1025–1032. [CrossRef]
10. Stibbe, D.T.; Tennyson, J. Near-threshold electron impact dissociation of $H_2$ within the adiabatic nuclei approximation. *New J. Phys.* **1998**, *1*, 2. [CrossRef]
11. Liu, X.; Shemansky, D.E.; Johnson, P.V.; Malone, C.P.; Khakoo, M.A.; Kanik, I. Electron and photon dissociation cross sections of the $H_2$ singlet ungerade continua. *J. Phys. B At. Mol. Opt. Phys.* **2012**, *45*, 105203. [CrossRef]
12. Stibbe, D.T.; Tennyson, J. Rates for the electron impact dissociation of molecular hydrogen. *Astrophys. J.* **1999**, *513*, 147–150. [CrossRef]
13. Tapley, J.K.; Scarlett, L.H.; Savage, J.S.; Fursa, D.V.; Zammit, M.C.; Bray, I. Electron-impact dissociative excitation of singlet states of molecular hydrogen. *Phys. Rev. A* **2018**, *98*, 032701. [CrossRef]

14. Scarlett, L.H.; Tapley, J.K.; Savage, J.S.; Fursa, D.V.; Zammit, M.C.; Bray, I. Vibrational excitation of the $H_2$ $X^1\Sigma_g^+$ state via electron-impact excitation and radiative cascade. *Plasma Sources Sci. Technol.* **2019**, *28*, 025004. [CrossRef]
15. Tapley, J.K.; Scarlett, L.H.; Savage, J.S.; Zammit, M.C.; Fursa, D.V.; Bray, I. Vibrationally resolved electron-impact excitation cross sections for singlet states of molecular hydrogen. *J. Phys. B At. Mol. Phys.* **2018**, *51*, 144007. [CrossRef]
16. Zammit, M.C.; Fursa, D.V.; Savage, J.S.; Bray, I. Electron– and positron–molecule scattering: Development of the molecular convergent close-coupling method. *J. Phys. B At. Mol. Opt. Phys.* **2017**, *50*, 123001. [CrossRef]
17. Lane, N.F. The theory of electron-molecule collisions. *Rev. Mod. Phys.* **1980**, *52*, 29–119. [CrossRef]
18. Glass-Maujean, M.; Liu, X.; Shemansky, D.E. Analysis of Electron-Impact Excitation and Emission of the $np\sigma^1\Sigma_u^+$ and $np\pi^1\Pi_u$ Rydberg Series of $H_2$. *Astrophys. J. Suppl. Ser.* **2009**, *180*, 38. [CrossRef]
19. Trevisan, C.S.; Tennyson, J. Differential cross sections for near-threshold electron impact dissociation of molecular hydrogen. *J. Phys. B At. Mol. Opt. Phys.* **2001**, *34*, 2935–2949. [CrossRef]
20. Trevisan, C.; Tennyson, J. Calculated rates for the electron impact dissociation of molecular hydrogen, deuterium and tritium. *Plasma Phys. Control. Fusion* **2002**, *44*, 1263–1276. [CrossRef]

atoms

MDPI

Article

# Charge Exchange Cross Sections for Noble Gas Ions and $N_2$ between 0.2 and 5.0 keV

**Steven Bromley *, Corey Ahl, Chad Sosolik and Joan Marler**

Department of Physics and Astronomy, Clemson University, Clemson, SC 29634, USA; cdahl@g.clemson.edu (C.A.); sosolik@clemson.edu (C.S.); jmarler@clemson.edu (J.M.)

* Correspondence: sjbroml@clemson.edu

Received: 30 June 2019; Accepted: 8 October 2019; Published: 14 October 2019

check for updates

**Abstract:** Charge transfer of an electron from a neutral atom to an ion is a fundamental interaction that plays a dominant role in the energy balance of atmospheric and astrophysical plasmas. The present investigation measured the charge exchange cross sections of noble gas ions ($He^+$, $Ne^+$, $Ar^+$, $Kr^+$) with $N_2$ in the intermediate energy range 0.2–5.0 keV. The systems were chosen because there remains a lack of consensus amongst previous measurements and regions where there were no previous measurements. A description of the mechanical design for an electrically floated gas cell is described herein.

**Keywords:** cross section; charge exchange; noble gas ions; $N_2$; molecules; scattering

## 1. Introduction

Charge exchange between ions and neutrals is an important process in a number of fields including fusion science [1], astrophysics [2–5], and atmospheric physics [6–8]. Recently, charge exchange between stellar wind ions ($H^+$, $He^+$, $He^{2+}$) and molecules has been investigated because of its relevance in astronomy e.g., with CO and $CO_2$ present in cometary atmospheres [9,10] and $N_2$ present in planetary atmospheres [11]. The charge exchange process is written

$$A^+ + B \rightarrow A + B^+ + \Delta E, \tag{1}$$

where ion $A^+$ carries the majority of the collision energy compared to neutral $B$. The change of internal energy as a result of the electron transfer is known as the energy defect $\Delta E$.

When $\Delta E = 0$, the process is resonant as in the case of symmetric collisions (i.e., A = B) where the cross section is well explained by semi-classical calculations over most energy ranges [12,13]. Asymmetric reactions (A $\neq$ B) may exhibit similar behavior for reactants with small energy defects, e.g., $H^+ + O \rightarrow H + O^+ + \Delta E(\sim 0)$ [7]. For other reactions, the energy defect is large ($\Delta E > 0$) and consequently the cross section is typically smaller [14]. However, cross sections for reactions with a large $\Delta E$ may become significant in certain high collision energy regions. When reactant B is a molecule, the number of possible reaction channels is large compared to the atomic case, and the energy dependence of the charge exchange cross section deviates from the single-peaked structure seen in asymmetric reactions such as those compiled by [15]. This is illustrated in particular by collisions between $N_2$ and atmospheric ions $H^+$ and $O^+$ in the summary of [7]. In the case of the $He^+$-$N_2$ reaction, discrepancies still remain between a variety of experimental efforts [11,16–21], including as much as a factor of three in magnitude and qualitatively different behaviors as a function of collision energy.

Charge exchange cross sections can be measured either with a gas cell apparatus or crossed-beam apparatus. While a crossed-beam set-up more easily allows for extraction of the target products, a gas cell allows for precise measurement of the target pressure and the effective path length. For a more in-depth discussion of these two techniques, see the review by [7]. Absolute cross sections are dependent on the ability to measure the ion beam currents and neutral target pressures accurately. The experiments performed herein took place in a gas cell apparatus [22] with the goal of producing absolute cross section measurements. The gas cell was designed with a small effective path length in order to measure charge exchange cross sections with large magnitudes ($10^{-15}$ cm$^2$) and mitigate the problem of large angle scattering.

In Section 2, we present new measurements for the charge exchange cross sections for the near-resonant reactions between noble gas ions and $N_2$ ($He^+$ $\Delta E = 9$ eV, $Ne^+$ $\Delta E = 6$ eV, $Ar^+$ $\Delta E = 0.2$ eV, $Kr^+$ $\Delta E = -1.6$ eV) for ion energies between 0.2 and 5.0 keV. In Section 3, we present details on the updated gas cell apparatus designed to provide better absolute pressure measurements in the gas cell region and improve data taking efficiency; we also discuss future directions.

## 2. Charge Exchange with $N_2$

### 2.1. Experimental Details

The gas cell (shown in Figure 1) is mounted on a 3D translational plus rotational ultra high vacuum (UHV) manipulator, with the intention of having the most flexibility to optimize the alignment with various ion beam sources. This so-called manipulator-mounted gas cell (MGC) includes two skimmers (1 mm and 2 mm apertures), a front end cap (3 mm aperture), a gas cell body (40 mm length), a back end cap (4 mm aperture), a retarding field analyzer (5 mm aperture), and a suppression electrode (6 mm aperture, −120 V). See [22] for more details.

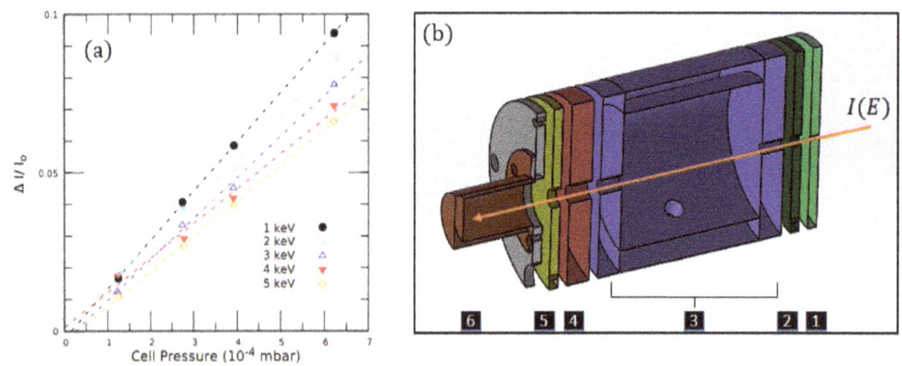


**Figure 1.** (**a**) Pressure-dependent current loss data obtained for the $Ar^+$ + Ar charge exchange process at incident ion energies of 1–5 keV. (**b**) Schematic of the gas cell used for charge exchange measurements: 1. Faceplate, 2. Skimmer, 3. Gas cell, 4. Retarding field analyzer, 5. Suppression electrode, 6. Faraday cup [22]. Reprinted from Ref. [Symmetric charge exchange for intermediate velocity noble gas projectiles. *J. Phys. B At. Mol. Opt. Phys.* **2019**, *52*, 215203], © IOP Publishing. Reproduced with permission. All rights reserved" with an appropriate reference to the other paper (https://doi.org/10.1088/1361-6455/ab42d1).

Beams of ions $A^+$ (A = He, Ne, Ar, Kr) are produced in an Omicron ISE 10 Sputter Ion Source. The gas cell is aligned to be colinear with the ion beam by finding the position which maximizes the current collected in the Faraday cup. The copper Faraday cup has an aspect ratio of 0.3 to minimize errors due to secondary electrons. The acceptance angle for detection is 2.0° for collisions occurring near the front of the cell and 13.0° at the back of the cell

The measurement procedure was performed as follows: The beam current $I_0(\epsilon)$ at each axial kinetic energy $\epsilon$ is maximized with an empty gas cell by scanning the focal and extraction electrodes on the source for each energy studied. Then, the gas cell is filled with the target gas. The target pressure is measured with a Bayard–Alpert gauge.

After the pressure stabilizes, $I(\epsilon)$, the beam current with gas in the cell is measured for each of the energy values by manually adjusting the energy, focus, and extraction voltages to the values previously determined. This is repeated for four different cell pressures spaced evenly between 1 and 8 $\times 10^{-4}$ mbar. The fractional current loss, $\frac{I_0(\epsilon)-I(\epsilon)}{I_0(\epsilon)}$, is plotted as a function of pressure for each $\epsilon$ and fitted to a linear function, as shown in Figure 1a. The charge exchange cross section is

$$\sigma_{\text{cx}}(\epsilon) = \frac{k_b T}{PL}\left(\frac{I_0(\epsilon) - I(\epsilon)}{I_0(\epsilon)}\right), \tag{2}$$

where $k_b$ is the Boltzmann constant, $T$ is the absolute temperature of the neutral gas (296 K), and $P$ is the pressure in the gas cell. $L$ is the effective path length of the interaction region (52.6 mm). In practice, $\sigma_{cx}$ is calculated from the slope, $m$, of the fractional current loss vs. pressure graph as follows:

$$\sigma_{\text{cx}}(\epsilon) = \frac{mk_b T}{L}. \tag{3}$$

All experiments were performed in the single collision regime ($P < 10^{-3}$ mbar) as determined by observing a strictly linear relationship between ion loss and pressure. We estimated the systematic uncertainty in the cross section at ±17% and observed statistical uncertainties on the order of ±10% (for more details see [22]).

### 2.2. $N_2$ Results

The charge exchange cross sections for the noble gas ions and neutral $N_2$ are shown in Figure 2a–d. The $He^+$+$N_2$ reaction has been studied across a wide energy range [11,16–21]. All seven previous results shown in Figure 2a utilized a gas cell or similar apparatus, and all but Rudd et al. [16] and Koopman [19] were able to measure both the charged and neutral components of the ion beam post-charge exchange (CX). By looking at the slow collision products, both Koopman (1968) [19] and Stebbings [20] identified the dominant CX channel as the disassociation $He^+ + N_2 \rightarrow He + N^+ + N$ via a two-step process.

As shown in Figure 2a, experiments at energies above 10 keV agree in trend, showing a peak at around 30 keV. For energies below 10 keV, the works of [17,19,20] show a monotonic decrease in the cross section with increasing energy. The most recent work [11] suggests a drop off toward low energies which is inconsistent with the $He^+$-$N_2$ results from [19,20]. The apparatus of Kusakabe et al. [11] is able to examine the charged and fast neutral components of their primary ion beam but are not equipped to detect the slow collision products. The measured decrease in $\sigma_{cx}$ with decreasing energy by [11] was argued to be a consequence of the energy defect for the dissociation process, $\Delta E \sim$ 0.2–0.3 eV, which suggests a maximum in the cross section of around 0.3 keV.

Our measurements show a monotonic decrease with increasing energy; however, our cross sections are systematically higher. A voltage of −120 V was applied to the suppression electrode to reflect any secondary electrons produced from the fast neutral component of the ion beam hitting the Faraday cup. Our measured cross sections include a contribution from elastic and inelastic scatter from angles above 13°.

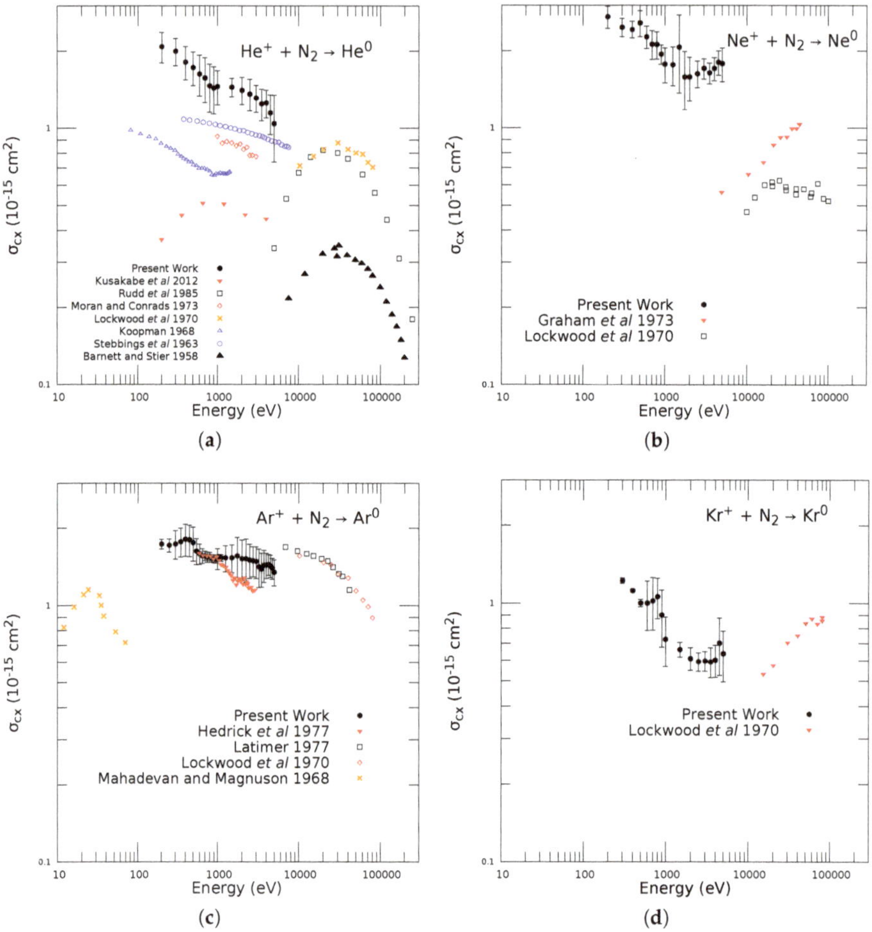


**Figure 2.** Total charge exchange cross sections for $A^+ + N_2 \rightarrow A$ for A = (He, Ne, Ar, Kr). The error bars shown represent statistical uncertainty in the measurements. (**a**) Total charge exchange cross sections for $He^+$ and $N_2$. The measurements shown: full circles (present work, manipulator-mounted gas cell (MGC)), inverted triangles [11], hollow squares [16], diamonds [17], × [18], hollow triangles [19], open circles [20], triangles [21]. (**b**) Total charge exchange cross sections for $Ne^+$ and $N_2$. The measurements shown: black circles (present work, MGC), inverted triangles [23], squares [18]. (**c**) Total charge exchange cross sections for $Ar^+$ and $N_2$. The measurements shown: black circles (present work, MGC), inverted triangles [24], squares [25], diamonds [18], × [26]. (**d**) Total charge exchange cross sections for $Kr^+$ and $N_2$. The measurements shown: black circles (present work, MGC), inverted triangles [18].

For $Ne^{+}+N_2$, only two experimental results are available [18,23]. Lockwood's cross sections [18] are nearly half an order of magnitude smaller than the present work and span an energy range outside the scope of our apparatus. The measurements by Graham et al. [23] studied the charged and neutral components of their ion beam post-CX, but their measured cross sections were placed on an absolute scale by comparison to known cross sections of protons in the respective target gases. The only point where the measurements of [23] overlap with the present work is at 5 keV where the present measurements are four times larger. As this reaction's angular dependence is unstudied, our data again include contributions from large angle elastic and/or inelastic scatter. Assuming the other works in Figure 2b isolated the charge exchange component of the total cross section, our measurements indicate that large angle non-CX scatter is comparable to CX between 0.2 and 5.0 keV.

$Ar^{+}+N_2$ collisions have been studied from 0.01 to 100 keV by several groups [18,24–26]. The earliest work by [26] measured the energy distribution of both the charged and neutral component of the transmitted ion beam, and inferred the dominant reaction channel at low energy was $Ar^{+} + N_2(v = 0) \rightarrow Ar + N_2^{+}(v = 1) + 0.093$ eV. At higher energies, [24] suggests that the large number of near-resonant reaction channels leads to a relatively flat cross section as a function of energy. For measurements between 0.5 and 1 keV, we find excellent agreement between the present work and [24]. The present work and [24] have the same magnitude cross section as the higher energy results of [18,25].

For $Kr^{+}$-$N_2$, only one previous work is available [18]. This paper contains measurements for all four systems studied here for energies above 10 keV. Unlike Lockwood's results in $He^{+}$ and $Ar^{+}$, their data show an increase in $\sigma_{cx}$ with increasing energy for $Kr^{+}$-$N_2$. This indicates that the collision energy of the cross section maximum is higher than in the other cases. We note that of the four reactions studied here, only the $Kr^{+}$-$N_2$ reaction contains a negative energy defect $\Delta E$.

## 3. Floating Gas Cell

Having the gas cell attached to the manipulator was very useful for aligning the beam and gas cell while the apparatus was under vacuum. However, it did add some constraints to the design, specifically the pressure gauge for the gas cell being located 14 inches from the cell and connected by 1/4 inch flexible stainless steel tubing. As a result of this, it took several minutes after each pressure change for equilibration to be reached between the gas cell and the gauge. With the current ion source, each adjustment of the beam energy requires manual resetting of the three ion source voltages. The combination of manually adjusting the source and waiting for the pressure to stabilize meant that a typical measurement of 10 different energies took in the order of 1 h. As a check of the stability of the ion source, the empty-cell beam currents were again measured after the gas was pumped out of the gas cell. If significant deviations between the pre- and post-gas currents were seen, the data was discarded and retaken. Lastly, the long tubing for gas injection led to 1–2 h returns to system base pressures which were in the $2 \times 10^{-8}$ mbar range.

A new gas cell was designed to improve the accuracy of the pressure measurements and allowed for faster data collection. A schematic of the electrically floated gas cell (FGC) is shown in Figure 3. The stainless steel gas cell body is mounted to an 8″ CF flange by a hollow (1.3″ internal diameter) rod. While no longer attached to the 3D manipulator for in-situ alignment, the design of the rod mount allows for the cell to be translated and rotated during installation. Alignment between the ion source exit and gas cell apertures is achieved with a HeNe laser. The hollow rod additionally provides a significantly larger conductance of the gas between the cell and the gauge. Not only is this an improvement in our ability to measure the pressure during experiments, the elimination of the narrow tubing of the MGC design results in a significant reduction in the time it takes to pump out the cell, and a reduction by a factor of four in system base pressure ($5 \times 10^{-9}$ mbar).

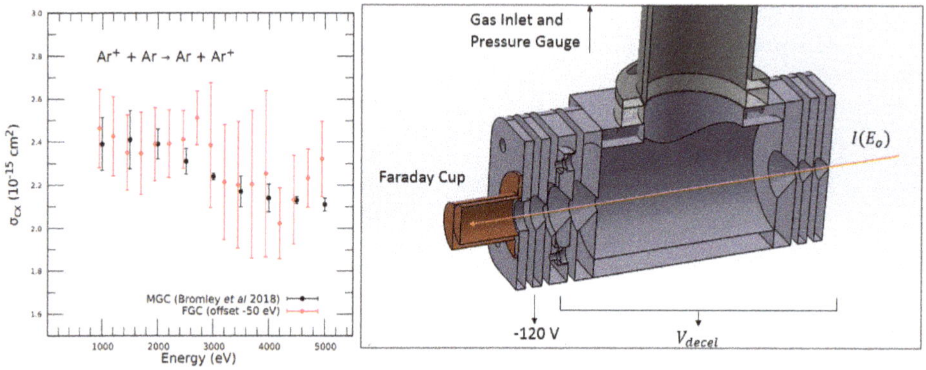


**Figure 3.** (**Left**) Comparison of $\sigma_{cx}$ for $Ar^+$ symmetric charge exchange measured with the floating gas cell (FGC) and the manipulator gas cell (MGC) from [22]. Data for the FGC is offset −50 eV for clarity. (**Right**) Cut-away view of the FGC. Shown is the gas inlet, MACOR spacer (mesh not shown), and gas cell components with knife edges. The direction of the incoming beam is indicated by the orange arrow.

The dimensions and placements of the electrical components of the FGC were the same as those in the MGC except as described below. Both the faceplate and skimmer were changed from the original aperture sizes (2, 3 mm) to (1, 1 mm), respectively. The FGC electrical components, including the two skimmers, end caps, retarding field analyzer, and suppression electrode, were machined with knife edges such that scattering off of the walls of the apertures is minimized. We also modified the retarding field analyzer by adding a 100×100 stainless steel mesh (wire thickness 0.001") in its aperture to ensure a uniform field. All components were electrically isolated with MACOR top-hat washers.

The main difference, however, is that the gas cell is electrically isolated from the gas inlet (and therefore the rest of the vacuum chamber) by a 1/4" thick MACOR ring. On the underside of the MACOR ring, a thin annular holder braces a titanium mesh (20 × 20, wire diameter 0.016") between the MACOR ring and cell body. Teflon sleeves insulate the screw heads between the cell body and metal gas inlet. With the FGC, rather than adjusting the beam energy $\epsilon$ at the ion source as in the MGC design, the FGC beam energy is adjusted by applying an electric potential $V_{\mathrm{FGC}}$ to all of the gas cell components (i.e., skimmers, end caps, cell body, and retarding field analyzer). This was done using a digitally controlled high voltage supply integrated into our system's LabVIEW control software. The beam energy in the cell is then given by $\epsilon = |e|(V_{source} - V_{\mathrm{FGC}})$.

The experimental procedure was the same as previously described except that the parameters of the ion beam source were fixed providing a single source energy throughout an experiment. After recording data for a series of beam energies and cell pressures, the analysis was done using Equation (3), the same as with the MGC data. In Figure 3, we can see a comparison of the measurements of the $Ar^+$-Ar cross section in the MGC [22] and preliminary tests using the FGC. For the $Ar^+$-Ar measurements from the MGC [22] in Figure 3, the statistical uncertainties are in the order of ±5%. In the FGC data, the statistical uncertainties in the cross sections are in the order of ±9%. The use of 1 mm apertures in both the faceplate and skimmer decrease the beam current significantly compared to the MGC design ($I_o^{MGC} \sim$ 10–60 nA, $I_o^{FGC} \sim$ 2–10 nA, depending on ion source pressure and beam energy), and the smaller beam currents lead to larger noise in the Faraday cup current measurements.

It is important to note that with the MGC design, one experimental set (consisting of 10 energies each) either spanning 0.2–1.0 keV or 1.0–5.0 keV, took 3 h due to the long pumping times (up to 2 h) and the measurement duration (1 h). With the FGC design, one experimental set, spanning 2 keV in $E$, may be accomplished in 0.5 h, and thus a single reaction in Figure 2 may be accomplished in 8 h (breaking the energy range in two sections and performing each four times). Increasing the ion source operating pressure and/or reverting the faceplate and skimmer apertures to their original sizes (2, 3 mm) would increase the measured beam currents and reduce the statistical uncertainties in the measured cross sections.

## 4. Conclusions

We measured total charge exchange cross sections for the reactions $A^+ + N_2 \rightarrow A$ for ions A = (He, Ne, Ar, Kr) between 0.2 and 5.0 keV. From comparisons with other measurements in $N_2$, our results suggest large-angle scattering may be important in ion-molecule scattering in the keV energy range. Presently, the MGC design cannot distinguish between current loss due to scatter beyond 13° and neutralization of ions from CX. Future experiments will investigate the effect of larger aperture sizes on the measured cross sections. Increasing the aperture sizes of the back end cap from 4 to 6 mm and the Faraday cup, suppression electrode, and retarding field analyzer to 9 mm will yield an increase in the collection angle from 2° (present) to 4° at the front of the cell and from 13° (present) to 18° at the rear of the cell. In total, this change will increase the effective collection area of our primary ion beam and help constrain the effect of scattering to large angles in these noble gas ion-molecule collisions.

The design and first measurements with an electrically floating gas cell were also presented. Preliminary results with this new design are in excellent agreement with previous measurements from our group, but also show larger statistical uncertainties. However, the faster measurement protocol means that we can compensate by taking more measurements. Other methods for increasing the signal to noise will include increasing the ion beam current density and/or working with larger aperture radii.

**Author Contributions:** C.S. and J.M. contributed to the design of the apparatus and oversaw the experimental program. C.A. designed the Floating Gas Cell. S.B. carried out construction and testing of the Floating Gas Cell. All experimental measurements and data analysis were carried out by S.B. All authors contributed to preparing this manuscript.

**Acknowledgments:** The authors gratefully acknowledge the work of the Physics and Astronomy Instrumentation Shop and financial support from the Clemson University College of Science.

**Conflicts of Interest:** The authors declare no conflict of interest.

## Abbreviations

The following abbreviations are used in this manuscript:

| | |
|---|---|
| CX | charge exchange |
| FGC | Floated Gas Cell |
| MGC | Manipulator Gas Cell |

## References

1. Janev, R.; Harrison, M.; Drawin, H. Atomic and molecular database for fusion plasma edge studies. *Nucl. Fusion* **1989**, *29*, 109. [CrossRef]
2. Schwadron, N.A.; Cravens, T.E. Implications of Solar Wind Composition for Cometary X-rays. *Astrophys. J.* **2000**, *544*, 558. [CrossRef]
3. Fabian, A.C.; Sanders, J.S.; Williams, R.J.R.; Lazarian, A.; Ferland, G.J.; Johnstone, R.M. The energy source of the filaments around the giant galaxy NGC 1275. *Mon. Not. R. Astron. Soc.* **2011**, *417*, 172–177. [CrossRef]
4. Bhardwaj, A.; Dhanya, M.B.; Alok, A.; Barabash, S.; Wieser, M.; Futaana, Y.; Wurz, P.; Vorburger, A.; Holmström, M.; Lue, C.; et al. A new view on the solar wind interaction with the Moon. *Geosci. Lett.* **2015**, *2*, 10. [CrossRef]

5. Mullen, P.D.; Cumbee, R.S.; Lyons, D.; Gu, L.; Kaastra, J.; Shelton, R.L.; Stancil, P.C. Line Ratios for Solar Wind Charge Exchange with Comets. *Astrophys. J.* **2017**, *844*, 7. [CrossRef]
6. Roble, R.G.; Ridley, C.C. An auroral model for the NCAR thermospheric general circulation model (TGCM). *Ann. Geophys. A* **1987**, *5*, 369–382.
7. Lindsay, B.G.; Stebbings, R.F. Charge transfer cross sections for energetic neutral atom data analysis. *J. Geophys. Res.* **2005**, *110*, A12213. [CrossRef]
8. Larsson, M.; Geppert, W.D.; Nyman, G. Ion chemistry in space. *Rep. Prog. Phys.* **2012**, *75*, 066901. [CrossRef]
9. Bodewits, D.; Hoekstra, R.; Seredyuk, B.; McCullough, R.W.; Jones, G.H.; Tielens, A.G.G.M. Charge Exchange Emission from Solar Wind Helium Ions. *Astrophys. J.* **2006**, *642*, 593. [CrossRef]
10. Werbowy, S.; Pranszke, B. Charge-exchange processes in collisions of $H^+$, $H_2^+$, $H_3^+$, $He^+$, and $He_2^+$ ions with CO and $CO_2$ molecules at energies below 1000 eV. *Phys. Rev. A* **2016**, *93*, 022713. [CrossRef]
11. Kusakabe, T.; Kitamuro, S.; Nakai, Y.; Tawara, H.; Sasao, M. Charge-Transfer Cross Sections of Ground State $He^+$ Ions in Collisions with He Atoms and Simple Molecules in the Energy Range below 4.0 keV. *Plasma Fusion Res.* **2012**, *7*, 2401062. [CrossRef]
12. Rapp, D.; Francis, D.E. Charge Exchange between Gaseous Ions and Atoms. *J. Chem. Phys.* **1962**, *37*, 2631–2645. [CrossRef]
13. Hodgkinson, D.P.; Briggs, J.S. Resonant charge exchange at low velocities. *J. Phys. B At. Mol. Phys.* **1976**, *9*, 255. [CrossRef]
14. Massey, H.S.W. Collisions between atoms and molecules at ordinary temperatures. *Rep. Prog. Phys.* **1949**, *12*, 248. [CrossRef]
15. Friedman, B.; DuCharme, G. Semi-empirical scaling for ion-atom single charge exchange cross sections in the intermediate velocity regime. *J. Phys. B At. Mol. Opt. Phys.* **2017**, *50*, 115202. [CrossRef]
16. Rudd, M.E.; Goffe, T.V.; Itoh, A. Cross sections for ionization of gases by 10–2000 keV $He^+$ ions and for electron capture and loss by 5–350 keV $He^+$ ions. *Phys. Rev. A* **1985**, *32*, 829–835. [CrossRef]
17. Moran, T.F.; Conrads, R.J. Charge neutralization of He+ ion beams. *J. Chem. Phys.* **1973**, *58*, 3793–3799. [CrossRef]
18. Lockwood, G.J. Total Cross Section for Charge Transfer of Noble-Gas Ions $N_2$. *Phys. Rev. A* **1970**, *2*, 1406–1410. [CrossRef]
19. Koopman, D.W. Light-Ion Charge Exchange in Atmospheric Gases. *Phys. Rev.* **1968**, *166*, 57–62. [CrossRef]
20. Stebbings, R.F.; Smith, A.C.H.; Ehrhardt, H. Dissociative Charge Transfer in He+–$O_2$ and He+–$N_2$ Collisions. *J. Chem. Phys.* **1963**, *39*, 968–971. [CrossRef]
21. Barnett, C.F.; Stier, P.M. Charge Exchange Cross Sections for Helium Ions in Gases. *Phys. Rev.* **1958**, *109*, 385–390. [CrossRef]
22. Bromley, S.J.; Fox, D.C.; Sosolik, C.E.; Harriss, J.E.; Marler, J.P. A gas cell apparatus for measuring charge exchange cross sections with multicharged ions. *Rev. Sci. Instrum.* **2018**, *89*, 073107. [CrossRef] [PubMed]
23. Graham, W.G.; Latimer, C.J.; Browning, R.; Gilbody, H.B. Ionization of fragmentation of diatomic molecular gases by 5–45 keV Ne+ and Na+ ions and Ne atoms. *J. Phys. B At. Mol. Phys.* **1973**, *6*, 2641–2652. [CrossRef]
24. Hedrick, A.F.; Moran, T.F.; McCann, K.J.; Flannery, M.R. Charge transfer cross sections in argon ion–diatomic molecule collisions. *J. Chem. Phys.* **1977**, *66*, 24–31. [CrossRef]
25. Latimer, C.J. Near-resonant charge transfer in collisions of rare-gas ions with simple molecules within the energy range 4–45 keV. *J. Phys. B At. Mol. Phys.* **1977**, *10*, 515–522. [CrossRef]
26. Mahadevan, P.; Magnuson, G.D. Low-Energy (1- to 100-eV) Charge-Transfer Cross-Section Measurements for Noble-Gas-Ion Collisions with Gases. *Phys. Rev.* **1968**, *171*, 103–109. [CrossRef]

atoms

MDPI

Article

# Quantemol Electron Collisions (QEC): An Enhanced Expert System for Performing Electron Molecule Collision Calculations Using the R-Matrix Method

**Bridgette Cooper [1], Maria Tudorovskaya [2], Sebastian Mohr [2], Aran O'Hare [1], Martin Hanicinec [1], Anna Dzarasova [2], Jimena D. Gorfinkiel [3], Jakub Benda [3], Zdeněk Mašín [4], Ahmed F. Al-Refaie [1], Peter J. Knowles [5] and Jonathan Tennyson [1,*]**

1 Department of Physics and Astronomy, University College London, London WC1E 6BT, UK; bridgette.cooper@ucl.ac.uk (B.C.); aran.ohare@ucl.ac.uk (A.O.); hanicinecm@gmail.com (M.H.); ahmed.al-refaie.12@ucl.ac.uk (A.F.A.-R.)

2 Quantemol Ltd., 320 City Rd, The Angel, London EC1V 2NZ, UK; mtudorov@quantemol.com (M.T.); s.mohr@quantemol.com (S.M.); adzarasova@quantemol.com (A.D.)

3 School of Physical Sciences, The Open University, Walton Hall, Milton Keynes MK7 6AA, UK; jimena.gorfinkiel@open.ac.uk (J.D.G.); jakub.benda@open.ac.uk (J.B.)

4 Institute of Theoretical Physics, Faculty of Mathematics and Physics, Charles University, V Holešovičkách 2, 180 00 Prague 8, Czech Republic; zde84nk@gmail.com

5 School of Chemistry, Cardiff University, Main Building, Park Place, Cardiff CF10 3AT, UK; knowlespj@cardiff.ac.uk

* Correspondence: j.tennyson@ucl.ac.uk

Received: 7 July 2019; Accepted: 8 October 2019; Published: 17 October 2019

check for updates

**Abstract:** Collisions of low energy electrons with molecules are important for understanding many aspects of the environment and technologies. Understanding the processes that occur in these types of collisions can give insights into plasma etching processes, edge effects in fusion plasmas, radiation damage to biological tissues and more. A radical update of the previous expert system for computing observables relevant to these processes, Quantemol-N, is presented. The new Quantemol Electron Collision (QEC) expert system simplifyies the user experience, improving reliability and implements new features. The QEC graphical user interface (GUI) interfaces the Molpro quantum chemistry package for molecular target setups, and the sophisticated UKRmol+ codes to generate accurate and reliable cross-sections. These include elastic cross-sections, super elastic cross-sections between excited states, electron impact dissociation, scattering reaction rates, dissociative electron attachment, differential cross-sections, momentum transfer cross-sections, ionization cross sections, and high energy electron scattering cross-sections. With this new interface we will be implementing dissociative recombination estimations, vibrational excitations for neutrals and ions, and effective core potentials in the near future.

**Keywords:** cross sections; elastic scattering; inelastic scattering; electronic excitation; rotational excitation; electron scattering; ionization; momentum transfer

## 1. Introduction

Electron collisions drive many processes both in the natural world and industrial processes. Compilations of electron collision cross sections, see for example [1–4], are becoming increasingly reliant on theory. Indeed the difficulty of measuring cross sections for chemically unstable species (radicals) has led to the suggestion that computed cross sections represent the solution for many problems involving technological plasmas [5,6]. Computed cross sections sets for key species are beginning to become available [7,8]. Computing such data sets, especially for low-energy collisions, requires sophisticated computer programs capable of treating both the electronic structure of the target molecule and the details of the electron collision process.

The R-matrix method [9] has proved very powerful for computing cross sections for a whole range of atomic and molecular problems [9]. For electron–molecule collisions the method has been implemented in a series of codes [10–13], generically known as the UK Molecular R-matrix (UKRmol) codes. These codes are powerful and can be used to tackle a variety of problems [14] but they are far from straightforward to run, as they require specialist knowledge on both molecular structure and electron scattering as well as experience on how to build sophisticated and balanced models, see [15–17] for example.

To address this problem Quantemol Ltd. developed an expert system, known as Quantemol-N (QN) [18], to run the UKRmol codes of Carr et al. [12]. Over time this code was developed to include not only the normal facilities of the UKRMol code to compute elastic and electronically inelastic cross sections but also to compute differential, momentum transfer and rotational excitation cross sections using code POLYDCS [19], ionization cross sections using the BEB (binary encounter Bethe) method of Kim and Rudd [20], and an especially developed dissociative electron attachment (DEA) estimator [21]. More recent developments involved the extension of calculations to high energies using the so-called BEf procedure of Kim [22] and the spherical complex optical potential (SCOP) method [23–25]. Recent releases of QN also feature an implementation of the UKRMol photoionization model [26].

Recently Benda, Mašín, Gorfinkiel and co-workers have developed a significantly updated and improved version of the UK Molecular R-matrix codes known as UKRmol+ [13]. New features in UKRMol+ include the implementation of new integrals package [27] which allows the use of Gaussian (GTO) [28] and B-spline (BTO) [29,30] type orbitals to represent the continuum. This new package facilitates the treatment of electron collisions with significantly larger molecules than was possible with UKRmol [31] and hence QN. UKRmol+ also features a greatly improved procedure for construction and diagonalization of the inner region Hamiltonian matrix [32], which is the rate-limiting step in most R-matrix calculations, as well as a number of other improvements.

Previous implementations of these codes essentially contained a quantum chemistry code at their heart. In order to construct the Hamiltonian matrix on the basis of configurations, a unique set of molecular orbitals describing the target molecule are needed. Previous implementations of the codes contained programs that allowed generation of these orbitals using simple approaches. UKRmol+ uses a different philosophy whereby the target molecular orbitals and the atomic GTO basis are read in from a file in the Molden format [33]. Although the Molden file is general and can be generated by a range of quantum chemistry software, most calculations actually use the general-purpose quantum chemistry program package Molpro [34,35]. This gives UKRmol+ access to orbitals generated using a wide range of sophisticated methods including state-averaged complete active space self-consistent field (CASSCF).

To take advantage of the improvements offered by the new UKRmol+ code, we decided to build a new expert system, known as Quantemol Electron Collisions or QEC for short. QEC incorporates all the features of QN discussed above plus new functionality made possible by UKRmol+ and closer integration with Molpro. The purpose of this paper is to describe the QEC expert system.

## 2. The R-Matrix Method

The R-matrix method is an example of an embedding method which divides space into different regions. In this case the method uses an inner region which is a sphere of radius $a$ and an outer region. In its most general form the inner region wave functions are represented by the close-coupling expansion:

$$\psi_k^{N+1} = \hat{A} \sum_{i,j} c_{ijk} \Phi_i^N(\mathbf{x}_1, \ldots, \mathbf{x}_N) \eta_{ij}(\mathbf{x}_{N+1}) + \sum_m b_{mk} \chi_m^{N+1}(\mathbf{x}_1, \ldots, \mathbf{x}_{N+1}). \tag{1}$$

In this equation $\Phi_i^N(\mathbf{x}_1, \ldots, \mathbf{x}_N)$ represents the target wave function. In QEC this wave function is generated using GTOs from the Molpro basis set library and orbital imported from Molpro generated with one of two models: either HF (Hartree–Fock) or MC-SCF (multi-configuration self consistent field) as discussed below. The $\eta_{ij}(\mathbf{x}_{N+1})$ are orbitals used to represent the continuum for which QEC has a library of GTO functions based on those of Faure et al. [28]. Functions with $\ell \leq 4$ (i.e., up to g wave) are explicitly included. The antisymmetriser, $\hat{A}$, ensures that the product of the target wave function and continuum orbital obeys the Pauli principle.

The second sum in Equation (1) runs over configurations so-called $L^2$ where all electrons are place in target orbitals. The precise configurations chosen depends on the model used for a particular calculation. QEC implements three distinct scattering models which are selected by the user.

Static exchange (SE) is the simplest model implemented; it uses HF target wave function and the $L^2$ configurations are given simply by placing the scattering electron in unoccupied target (virtual) orbitals.

Static exchange plus polarization (SEP) builds on the SE model by also including $L^2$ configurations which involve promoting an electron from the HF target wave function to a virtual orbital while also placing the scattering electron in a target virtual orbital. This provides a good representation of low-lying resonances which are the route to DEA. The SEP model is therefore recommended for computing DEA cross sections.

Close-coupling (CC) expansions involve including several target states in the inner region expansion, Equation (1). This model normally uses a complete active spaces (CAS) description of these states [15] for which it employs MCSCF orbitals (this is an improvment of QN which used HF orbitals for this step). Studies of electron-impact electronic excitation and dissociation (which goes via electronic excitation) must be performed at the CC level; this is also recommended for studies of Feshbach resonances. CC calculations are computationally more demanding than SE or SEP ones.

There is a fourth possible model for studying electron molecule collisions with UKRMol/UKRmol+, which is the R-matrix with pseudostates (RMPS) method [36,37]. The RMPS method is a generalization of the CC method which allows for the inclusion of target continuum states in the CC expansion. RMPS calculations have very useful properties in terms of extending the energy range of the calculations and converging polarization effects [38], but are computationally very demanding [16,39] so as yet have not been implemented in QEC.

Solving the problem in the outer region is performed in two steps. First the R-matrix is propagated from the boundary given by $r = a$ to some large $r = r_f$ [40], where $r_f$ is chosen such that the non-Coulombic potential can be neglected for asymptotic region defined by $r > r_f$; QEC assumes $r_f = 100.1$ a$_0$ and beyond this uses an asymptotic expansion due to Gailitis [41,42]. The K-matrices so calculated are used by QEC to automatically detect and fit resonances [43] and to construct the corresponding T-matrices. The resonance parameters form the main input to the DEA estimator [21]. The T-matrices are used to compute elastic and electronically inelastic cross sections, and provide the input to code POLYDCS [19] which computes differential, momentum transfer and rotational-excitation cross sections. For molecules with a permanent dipole moment it is necessary to correct for partial waves higher than $\ell = 4$ [44]; this done either using POLYDCS or code BornCross [45]. Finally, the elastic and electronically inelastic cross sections are turned into rates assuming Maxwellian distribution for the electron energies.

## 3. The QEC Interface

QEC runs through a graphical user interface (GUI) which guides the user through the calculations. The first inputs come on a screen which provides the molecule set up, see Figure 1. The geometry of the molecule can be put in by hand or via an xyz file. If the precise geometric parameters are not known, an approximate geometry can be guessed and is automatically optimized by Molpro. This is requested using the "Optimise" tab and results in a rapid Hartree–Fock level calculation. The figure illustrates the geometry of methane which has been optimized in this fashion.

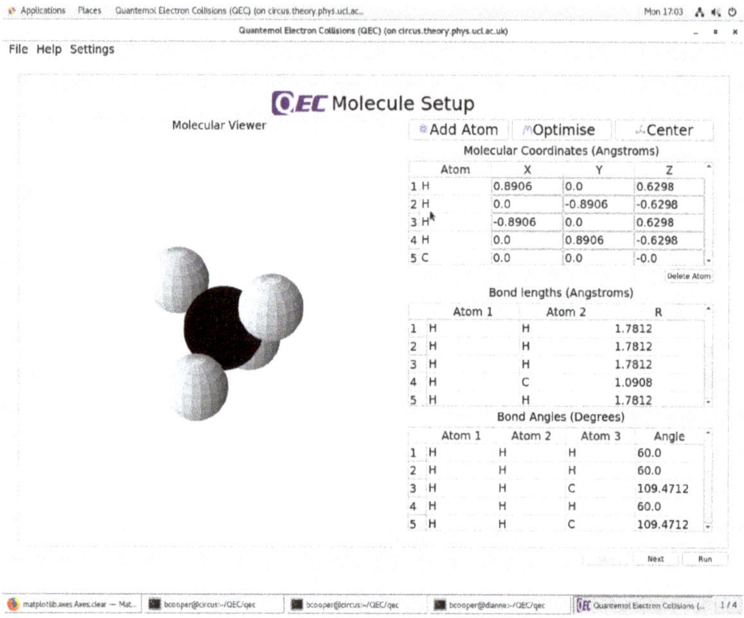


**Figure 1.** Screenshot of the molecular set-up page in the Quantemol Electron Collision (QEC) graphical user interface (GUI) showing methane as an example. The screenshot was taken after the geometry had been optimized using Molpro and the coordinates automatically shifted to the center of mass.

It is a requirement of the R-matrix method that the origin of the coordinate system is the molecular center of mass. QEC automatically shifts the geometry to this center of mass; it stores a library of main-isotope atomic masses for this purpose. Note that for molecular ions, the choice of center of mass is important for obtaining the correct target dipole moment [46]; the effect of isotopic substitution on the center of mass can be controlled via an advanced option.

Both Molpro and QEC only use Abelian symmetry groups i.e., ones which contain no degenerate representations. Molpro contains the facility to automatically classify molecules to a point group. QEC exploits this facility once the geometry has been determined. The resulting point group is illustrated using the appropriate mirror planes in the QEC molecular viewer which forms part of second input page of the GUI. Figure 2 shows this for the example of methane. Note that while methane in its equilibrium geometry has point-group symmetry $T_d$, Molpro and QEC only use the $C_{2v}$ subgroup.

The second input page also allows the user to choose a target basis set from a number of standard GTO sets. The level of calculation used to generate the target orbitals and wave functions must be chosen at this point: HF orbitals are used for SE and SEP calculations, while MCSCF orbitals are required for CC calculations. Here, and elsewhere, there are a number of advanced user options which allow the experienced user to change the standard QEC defaults. Figure 2 shows the choice of MCSCF orbitals computed using a standard cc-pVDZ basis set. The advanced user options displayed are actually the default ones. However, we note that calculations on ionic targets require the charge of the target molecule to be selected as non-zero at this point. The default active space used for the MCSCF calculations is the standard one used by Molpro; it is the valence space, comprising all the orbitals of the highest $n$ quantum number that are occupied, that are active,so for methane that means 2s2p for C and 1s for H, with the 1s orbital of C treated as a core orbital. For large molecules with many active electrons it may be necessary to reduce this space.

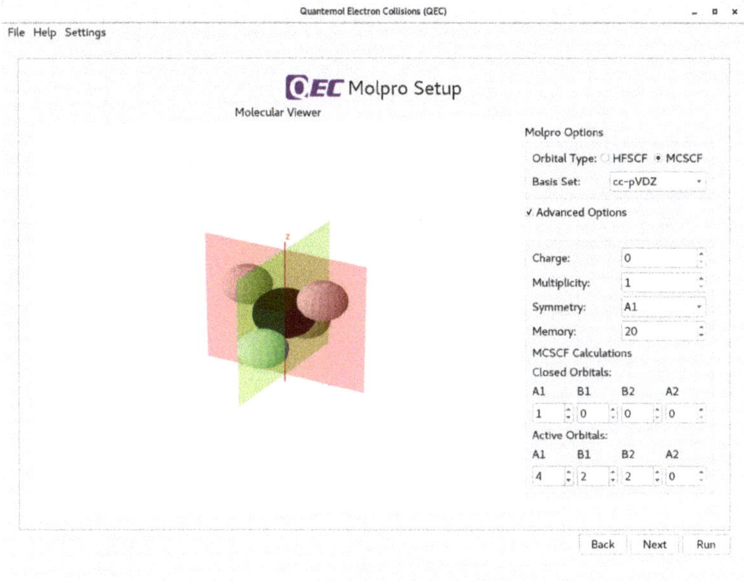


**Figure 2.** Screenshot of the page in the QEC GUI which sets up the molecular target calculation which is performed using Molpro.

Input page 3 of the GUI sets up the R-matrix and parameters for scattering calculation. The methane example shown in Figure 3 uses QEC defaults. At this stage the parameters of the calculations are now fully determined.

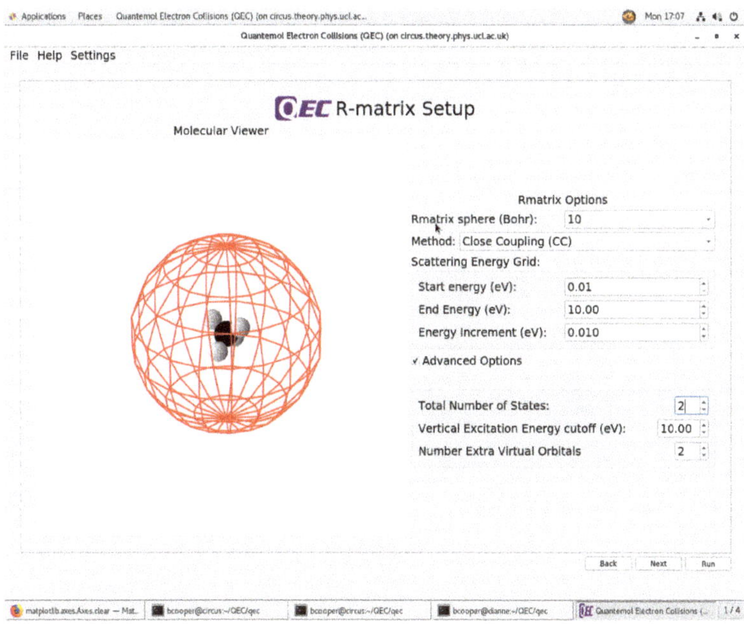


**Figure 3.** Screenshot of the page in the QEC GUI which set the parameters for the scattering calculation.

The final input page of the GUI, see Figure 4, allows the user to specify what results are required. Only outputs consistent with the chosen model (SE, SEP or CC) are allowed. Optionally the user can choose to compute differential, momentum transfer and rotational-excitation cross sections using code POLYDCS. Since POLYDCS only allows for electronically-elastic cross sections, these cross sections are only computed up to the first electronic excitation threshold in CC calculations. Choice of the DEA estimator requires the user to provide further information such as an estimated vibrational frequency of the dissociating mode and an estimated dissociation energy. We plan to use Molpro to provide these data automatically in the future. QEC already contains a library of electron affinities for standard atomic and molecular fragments.

The final two optional cross sections are the SCOP and BEf high energy approximations. SCOP calculations provide high energy estimates of the total elastic and inelastic (including ionization, etc.) cross section. SCOP calculations employ charge distributions provided by Molpro but are rather slow so should only be used if the user actually wants the results. Finally the BEf method is used to estimate high energy electronic excitation cross sections. The BEf method works using a dipole approximation [22] and only dipole-allowed electronic excitations are actually considered; this is consistent with the fact that excitations cross sections which are not dipole allowed go rapidly to zero at higher energies. Parameters for both BEB (ionization) and BEf calculations are fully provided by Molpro. For BEB the user has the option to change the ionization potential, which Molpro determines using Koopman's theorem, if they wish. Note that selecting BEB only at this stage will run only a BEB calculation which is very quick even for large systems.

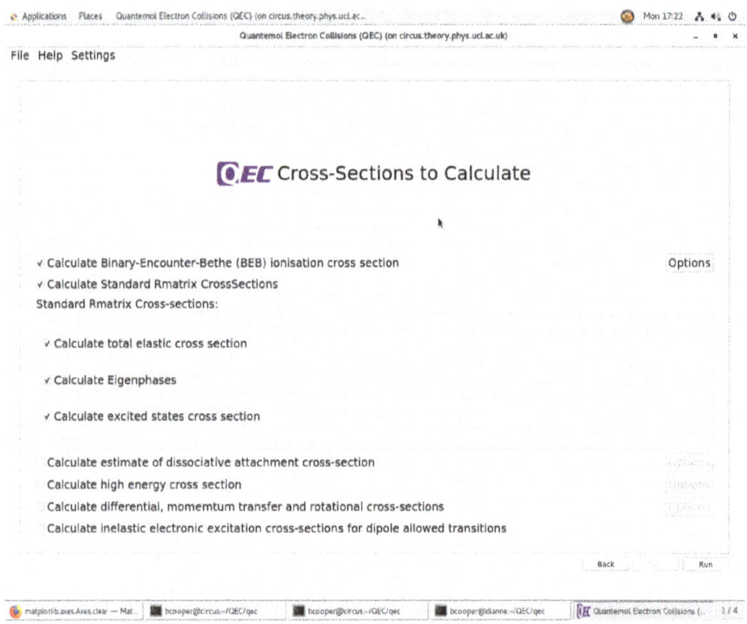


**Figure 4.** Sample screen for a close-coupling (CC) calculation on methane showing the outputs selected to be total cross section, the eigenphase sums for each symmetry, the rate of elastic scattering as a function of temperature and the ionization cross section computed using the binary encounter Bethe (BEB) method.

QEC provides a variety of results. Figure 5 illustrates some of the graphs automatically generated for the methane example considered above. Note that the $^2B_1$ and $^2B_2$ eigenphases are degenerate which is a usual occurrence when $C_{2v}$ symmetry is used as a proxy for a higher-symmetry point group. These eigenphases have been smoothed to remove arbitrary jumps by $\pi$. Note that the illustrated results for methane use a very simple SE model; a comprehensive R-matrix study of this problem is available [47].

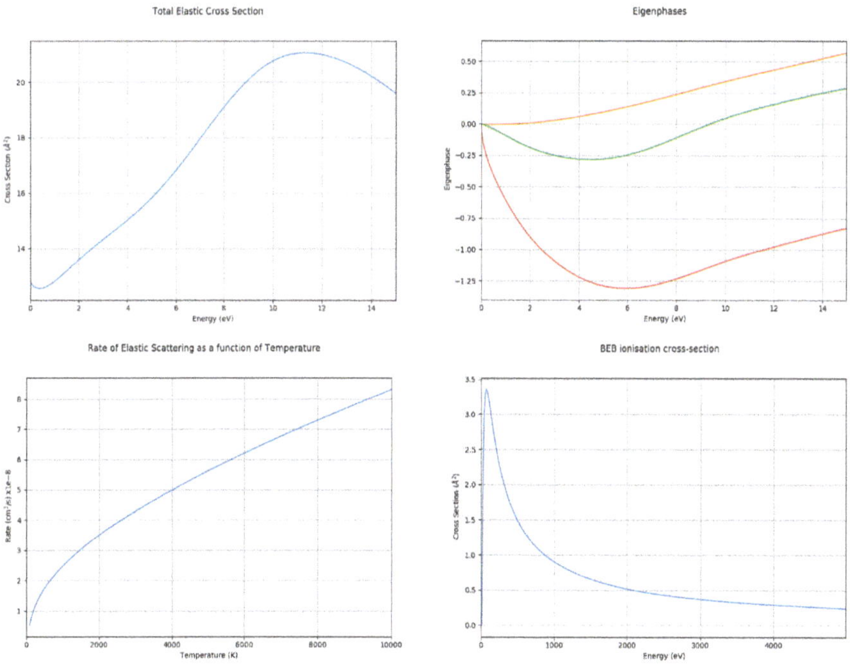


**Figure 5.** Examples of graphical outputs produced by QEC for a static exchange (SE) calculation on methane showing total cross sections, eigenphase sums, rates and ionization (BEB) cross sections.

Finally Figure 6 shows examples of plots displaying differential cross sections for $CO_2$ as a function of both electron collision energy and angle. The results are for an SEP calculation with eight virtual orbitals, using a cc-pVDZ target GTO basis. The results are plotted as single angle–energy dependent contour plots or, alternatively, as cuts through the DCS at a single angle or energy. Illustrated are the cross sections as a function of energy for a scattering angle of 126°. Figure 7 shows examples plots displaying differential cross sections for CO.

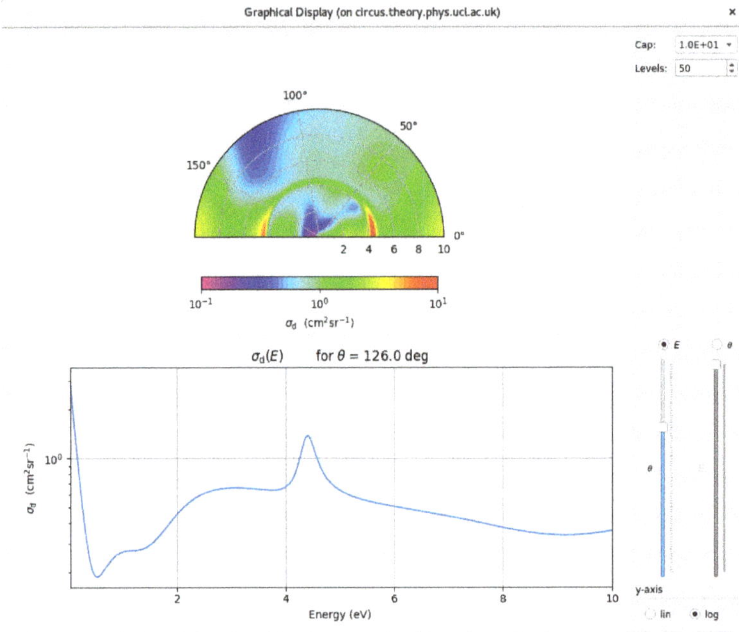


**Figure 6.** Differential cross sections for $CO_2$ as displayed by the QEC GUI.

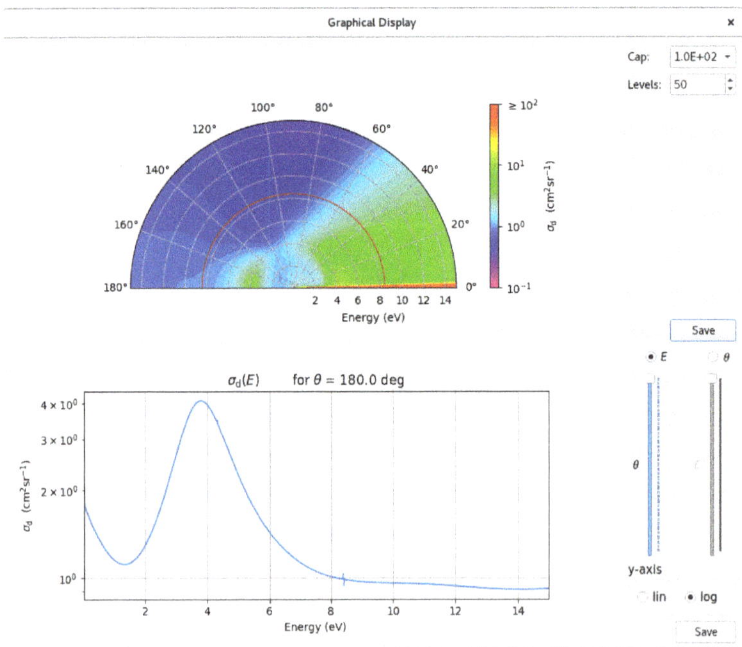


**Figure 7.** Differential cross sections for CO as displayed by the QEC GUI.

All the graphical results are also available as data spreadsheets. QEC also provides parameters for the resonances it detects and a theory model file gives details of the calculations performed plus appropriate references to quote when preparing the results for publication.

QEC is provided with a comprehensive manual and associated help facility. It also has a library of about a dozen previously run example which users can try and, if appropriate, adapt; this library is being systematically added to.

## 4. Concluding Remarks

The present article describes a new expert system designed for an easy and rapid generation of low-energy electron scattering data using the Quantemol Electron Collision (QEC) expert system. QEC runs an integrated versions of the Molpro ab initio molecular electronic structure code [34,35] and the newly developed UKRmol+ R-matrix electron molecule scattering code [13]. QEC exploits this close coupling of the codes in a number of ways such as automated geometry optimization and symmetry determination, and the direct provision of inputs for SCOP, BEB and BEf calculations. However there are a number of other places where this coupling will be exploited further in the future such as the automated provision of parameters for the DEA estimator.

Future releases of QEC will further increase its functionality. Plans include the implementation of a dissociative recombination estimator [48], automated electron-impact vibrational calculations, and the use of effective core potentials for heavy atoms.

**Author Contributions:** B.C. and M.T. were lead developers for QEC and were assisted by S.M., A.O., M.H., A.D. and J.T.; J.D.G., J.B., Z.M., A.F.A.-R. and J.T. developed UKRmol+; P.J.K. advised on the Molpro integration; J.T. and B.C. drafted the paper which was read and approved by all co-authors.

**Funding:** This work funded by UK research councils STFC and EPSRC grants numbers ST/R005133/1, EP/R029342/1, EP/G055556/1, EP/G055599/1, EP/P022146/1, EP/N509577/1 and EP/R513143/1, as well as eCSE projects eCSE01-13 and eCSE08-7.

**Conflicts of Interest:** QEC is distributed by Quantemol Ltd; Maria Tudorovskaya and Sebastian Mohr work for Quantemol while Anna Dzarasova and Jonathan Tennyson are company Directors. Aran O'Hare and Martin Hanicinec receive partial support for their studentships from Quantemol. Peter Knowles acts as a consultant to TTI GmbH, the owner of Molpro. None of the other authors are conflicted.

## References

1. Song, M.Y.; Yoon, J.S.; Cho, H.; Itikawa, Y.; Karwasz, G.P.; Kokoouline, V.; Nakamura, Y.; Tennyson, J. Cross sections for electron collisions with methane. *J. Phys. Chem. Ref. Data* **2015**, *44*, 023101. [CrossRef]
2. Song, M.Y.; Yoon, J.S.; Cho, H.; Karwasz, G.P.; Kokoouline, V.; Nakamura, Y.; Tennyson, J. Cross sections for electron collisions with acetylene. *J. Phys. Chem. Ref. Data* **2017**, *46*, 013106. [CrossRef]
3. Song, M.Y.; Yoon, J.S.; Cho, H.; Karwasz, G.P.; Kokoouline, V.; Nakamura, Y.; Hamilton, J.R.; Tennyson, J. Cross sections for electron collisions with $NF_3$. *J. Phys. Chem. Ref. Data* **2017**, *46*, 043104. [CrossRef]
4. Song, M.Y.; Yoon, J.S.; Cho, H.; Karwasz, G.P.; Kokoouline, V.; Nakamura, Y.; Tennyson, J. Electron collision cross sections with NO, $N_2O$ and $NO_2$. *J. Phys. Chem. Ref. Data* **2019**, in press.
5. Winstead, C.; McKoy, V. Electron-molecule collisions in low-temperature plasmas—The role of theory. *Adv. At. Mol. Phys.* **2000**, *43*, 111–145.
6. Bartschat, K.; Kushner, M.J. Electron collisions with atoms, ions, molecules, and surfaces: Fundamental science empowering advances in technology. *Proc. Natl. Acad. Sci. USA* **2016**, *113*, 7026–7034. [CrossRef] [PubMed]
7. Blanco, F.; Roldan, A.M.; Krupa, K.; McEachran, R.P.; White, R.D.; Marjanovic, S.; Petrovic, Z.L.; Brunger, M.J.; Machacek, J.R.; Buckman, S.J.; et al. Scattering data for modelling positron tracks in gaseous and liquid water. *J. Phys. B At. Mol. Opt. Phys.* **2016**, *49*, 145001. [CrossRef]
8. Hamilton, J.R.; Tennyson, J.; Huang, S.; Kushner, M.J. Calculated cross sections for electron collisions with $NF_3$, $NF_2$ and NF with applications to remote plasma sources. *Plasma Sources Sci. Technol.* **2017**, *26*, 065010. [CrossRef]

9. Burke, P.G. *R-Matrix Theory of Atomic Collisions: Application to Atomic, Molecular and Optical Processes*; Springer: New York, NY, USA, 2011.
10. Gillan, C.J.; Tennyson, J.; Burke, P.G. The UK molecular R-matrix scattering package: A computational perspective. In *Computational Methods for Electron-Molecule Collisions*; Huo, W., Gianturco, F.A., Eds.; Plenum: New York, NY, USA, 1995; pp. 239–254.
11. Gillan, C.J.; Tennyson, J.; McLaughlin, B.M.; Burke, P.G. Low energy electron impact excitation of the nitrogen molecule: optically forbidden transitions. *J. Phys. B At. Mol. Opt. Phys.* **1996**, *29*, 1531–1547. [CrossRef]
12. Carr, J.M.; Galiatsatos, P.G.; Gorfinkiel, J.D.; Harvey, A.G.; Lysaght, M.A.; Madden, D.; Mašín, Z.; Plummer, M.; Tennyson, J. The UKRmol program suite. *Eur. Phys. J. D* **2012**, *66*, 58. [CrossRef]
13. Benda, J.; Mašín, Z.; Gorfinkiel, J.D.; Harvey, A.G.; Tennyson, J. UKRmol+: A suite for modelling of electronic processes in molecules interacting with electrons, positrons and photons using the R-matrix method. *Comput. Phys. Commun.* **2019**, in press.
14. Tennyson, J. Electron-molecule collision calculations using the R-matrix method. *Phys. Rep.* **2010**, *491*, 29–76. [CrossRef]
15. Tennyson, J. R-matrix calculation of Rydberg states of CO. *J. Phys. B At. Mol. Opt. Phys.* **1996**, *29*, 6185–6201. [CrossRef]
16. Halmová, G.; Gorfinkiel, J.D.; Tennyson, J. Low and intermediate energy electron collisions with the $C_2^-$ molecular anion. *J. Phys. B At. Mol. Opt. Phys.* **2008**, *41*, 155201. [CrossRef]
17. Little, D.A.; Tennyson, J. Singlet and triplet ab initio Rydberg states of $N_2$. *J. Phys. B At. Mol. Opt. Phys.* **2013**, *46* 145102. [CrossRef]
18. Tennyson, J.; Brown, D.B.; Munro, J.J.; Rozum, I.; Varambhia, H.N.; Vinci, N. Quantemol-N: An expert system for performing electron molecule collision calculations using the R-matrix method. *J. Phys. Conf. Ser.* **2007**, *86*, 012001. [CrossRef]
19. Sanna, N.; Gianturco, F.A. Differential cross sections for electron/positron scattering for polyatomic molecules. *Comput. Phys. Commun.* **1998**, *114*, 142–167. [CrossRef]
20. Kim, Y.K.; Rudd, M.E. Binary-encounter-dipole model for electron-impact ionization. *Phys. Rev. A* **1994**, *50*, 3945. [CrossRef]
21. Munro, J.J.; Harrison, S.; Tennyson, J.; Fujimoto, M.M. A dissociative electron attachment cross-section estimator. *J. Phys. Conf. Ser.* **2012**, *388*, 012013. [CrossRef]
22. Kim, Y.K. Scaling of plane-wave Born cross sections for electron-impact excitation of neutral atoms. *Phys. Rev. A* **2001**, *64*, 032713. [CrossRef]
23. Calogero, F. *Variable Phase Approach to Potential Scattering*; Academic: New York, NY, USA, 1954.
24. Jain, A. Elastic scattering of electrons and positrons by $CH_4$ at 25–800 eV. *J. Chem. Phys.* **1983**, *78*, 6579–6583. [CrossRef]
25. Jain, A.; Baluja, K.L. Total (elastic plus inelastic) cross sections for electron scattering from diatomic and polyatomic molecules at 10–5000 eV: $H_2$, $Li_2$, HF, $CH_4$, $N_2$, CO, $C_2H_2$, HCN, $O_2$, HCl, $H_2S$, $PH_3$, $SiH_4$, and $CO_2$. *Phys. Rev. A* **1992**, *45*, 202–218. [CrossRef] [PubMed]
26. Brigg, W.J.; Harvey, A.G.; Dzarasova, A.; Mohr, S.; Brambila, D.S.; Morales, F.; Smirnova, O.; Tennyson, J. Calculated photoionization cross sections using Quantemol-N. *Jpn. J. Appl. Phys.* **2015**, *54*, 06GA02. [CrossRef]
27. Mašín, Z.; Benda, J.; Gorfinkiel, J.D. GBTOlib: A library for evaluation of molecular integrals in the basis of multicentric Gaussian and single-centre B-spline orbitals. In preparation.
28. Faure, A.; Gorfinkiel, J.D.; Morgan, L.A.; Tennyson, J. GTOBAS for fitting Gaussian Type Orbitals to Bessel and Coulomb functions. *Comput. Phys. Commun.* **2002**, *144*, 224–241. [CrossRef]
29. Bachau, H.; Cormier, E.; Decleva, P.; Hansen, J.E.; Martin, F. Applications of B-splines in atomic and molecular physics. *Rep. Prog. Phys.* **2001**, *64*, 1815–1943. [CrossRef]
30. Darby-Lewis, D.; Mašín, Z.; Tennyson, J. R-Matrix Calculations of electron-impact electronic excitation of BeH. *J. Phys. B At. Mol. Opt. Phys.* **2017**, *50*, 175201. [CrossRef]
31. Loupas, A.; Gorfinkiel, J.D. Shape and core-excited resonances in electron scattering from alanine. *J. Chem. Phys.* **2019**, *150*, 064307. [CrossRef]
32. Al-Refaie, A.F.; Tennyson, J. A parallel algorithm for Hamiltonian matrix construction in electron-molecule collision calculations: MPI-SCATCI. *Comput. Phys. Commun.* **2017**, *214*, 216–224. [CrossRef]

33. Schaftenaar, G.; Noordik, J.H. Molden: A pre- and post-processing program for molecular and electronic structures. *J. Comput.-Aided Mol. Des.* **2000**, *14*, 123–134. [CrossRef]
34. Werner, H.J.; Knowles, P.J.; Knizia, G.; Manby, F.R.; Schütz, M. Molpro: A general-purpose quantum chemistry program package. *WIREs Comput. Mol. Sci.* **2012**, *2*, 242–253. [CrossRef]
35. Werner, H.J.; Knowles, P.J.; Knizia, G.; Manby, F.R.; Schütz, M.; Celani, P.; Györffy, W.; Kats, D.; Korona, T.; Lindh, R.; et al. MOLPRO, Version 2019.2, a Package of ab Initio Programs, 2019. Available online: https://www.molpro.net/ (accessed on 14 October 2019).
36. Gorfinkiel, J.D.; Tennyson, J. Electron-$H_3^+$ collisions at intermediate energies. *J. Phys. B At. Mol. Opt. Phys.* **2004**, *37*, L343–L350. [CrossRef]
37. Gorfinkiel, J.D.; Tennyson, J. Electron impact ionisation of small molecules at intermediate energies: The R-matrix with pseudostates method. *J. Phys. B At. Mol. Opt. Phys.* **2005**, *38*, 1607–1622. [CrossRef]
38. Jones, M.; Tennyson, J. On the use of pseudostates to calculate molecular polarizabilities. *J. Phys. B At. Mol. Opt. Phys.* **2010**, *43*, 045101. [CrossRef]
39. Zhang, R.; Galiatsatos, P.G.; Tennyson, J. Positron collisions with acetylene calculated using the R-matrix with pseudo-states method. *J. Phys. B At. Mol. Opt. Phys.* **2011**, *44*, 195203. [CrossRef]
40. Morgan, L.A. A generalized R-matrix propagation program for solving coupled 2nd-order differential-equations. *Comput. Phys. Commun.* **1984**, *31*, 419–422. [CrossRef]
41. Gailitis, M. New forms of asymptotic expansions for wavefunctions of charged-particle scattering. *J. Phys. B At. Mol. Opt. Phys.* **1976**, *9*, 843. [CrossRef]
42. Noble, C.J.; Nesbet, R.K. CFASYM, a program for the calculation of the asymptotic solutions of the coupled equations of electron collision-theory. *Comput. Phys. Commun.* **1984**, *33*, 399. [CrossRef]
43. Tennyson, J.; Noble, C.J. RESON: For the automatic detection and fitting of Breit-Wigner resonances. *Comput. Phys. Commun.* **1984**, *33*, 421–424. [CrossRef]
44. Norcross, D.W.; Padial, N.T. The Multipole-Extracted Adiabatic-Nuclei Approximation for Electron-Molecule collisions. *Phys. Rev. A* **1982**, *25*, 226–338. [CrossRef]
45. Baluja, K.L.; Mason, N.J.; Morgan, L.A.; Tennyson, J. Electron scattering from ClO using the R-matrix method. *J. Phys. B At. Mol. Opt. Phys.* **2000**, *33*, L677–L684. [CrossRef]
46. Hamilton, J.R.; Faure, A.; Tennyson, J. Electron-impact excitation of diatomic hydride cations: $HeH^+$, $CH^+$, $ArH^+$. *Mon. Not. R. Astron. Soc.* **2016**, *455*, 3281–3287. [CrossRef]
47. Brigg, W.J.; Tennyson, J.; Plummer, M. R-Matrix calculations of low-energy electron collisions with Methane. *J. Phys. B At. Mol. Opt. Phys.* **2014**, *47*, 185203. [CrossRef]
48. Fonseca dos Santos, S.; Douguet, N.; Kokoouline, V.; Orel, A.E. Scattering matrix approach to the dissociative recombination of $HCO^+$ and $N_2H^+$. *J. Chem. Phys.* **2014**, *140*, 164308. [CrossRef] [PubMed]

atoms

MDPI

Article

# Improvement of the NIFS Atom and Molecular Database

**Masahiko Emoto [1,*], Izumi Murakami [1], Daiji Kato [1,2], Masanobu Yoshida [1], Masatoshi Kato [1] and Setsuo Imazu [1]**

1 National Institute for Fusion Science, Toki, Gifu 509-5202, Japan
2 Department of Advanced Energy Engineering Science, Kyushu University, Kasuga, Fukuoka 816-8580, Japan
* Correspondence: emoto.masahiko@nifs.ac.jp

Received: 29 June 2019; Accepted: 27 August 2019; Published: 11 September 2019

**Abstract:** The NIFS (National Institute for Fusion Science) Atom and Molecular Database, which has been available online since 1997, is a numerical atomic and molecular database of collision processes that is important for fusion research. This database provides the following: (1) the cross-sections and rate coefficients for ionization, excitation, and recombination caused by electron impact; (2) the charge transfer caused by heavy particle collision and collision processes of molecules; and (3) the sputtering yields of solids and backscattering coefficients from solids. It also offers a bibliographic database. We recently reconstructed the database system. The main purpose of the reconstruction was to migrate the database into an open-source architecture to make the system more flexible and extensible. The previous system used proprietary software and was difficult to customize. The new system consists of open-source software, including PostgreSQL database and Ruby on Rails. New features were also added to the system. The most important improvement is the interface with the Virtual Atomic and Molecular Data Center (VAMDC) portal. Using this interface, researchers can search for data in the NIFS database as well as in various other online databases simultaneously.

**Keywords:** databases; atoms; molecules; VAMDC; web

## 1. Introduction

Since 1975, atomic data have been compiled and published by a working group of atomic and plasma physics researchers from Japanese universities [1,2]. In the 1980s, these data were housed in an atomic database whose data retrieval and display systems were accessible via a mainframe computer system in the Institute of Plasma Physics at Nagoya University [3]. In 1997, the database system was migrated to a relational database, Oracle. The system was also uploaded online and named the NIFS Atom and Molecular Database (NIFS DB), enabling online data retrieval and display [4,5]. However, because of the rapid progress of computer and internet technology, several of the NIFS DB's software components had become obsolete and could not be updated further. Moreover, there was a potential security risk. Therefore, we decided to upgrade the system.

Another motivation for the upgrade was to allow the NIFS DB to be used by a wider range of scientists. The NIFS DB was originally constructed to gather information on electron impact ionization and excitation cross-sections and was later expanded to cover a wide variety of collision processes in plasma and plasma wall interactions. Table 1 shows the available contents of the data as of June 2019. Thus, the NIFS DB was produced mainly for fusion plasma research, but it is also useful in other areas of research, such as in astrophysics.

The Virtual Atomic and Molecular Data Center (VAMDC) Consortium runs an interoperable e-infrastructure to integrate a broad range of online atomic and modular databases. Through the VAMDC portal, a user can cross-search all the heterogeneous databases in the system [6]. By joining this integrated system, the NIFS DB will be widely accessible to other research groups beyond the plasma fusion research community. In order to join the VAMDC system, databases must conform to VAMDC standards, and the NIFS DB needs to be modified accordingly.

**Table 1.** Contents of the NIFS Atom and Molecular Database (NIFS DB) as of June 2019. The NIFS DB stores numerical and bibliographic data.

| Name | | Description | Records |
|---|---|---|---|
| *AMIDIS* | | Cross-sections for ionization, excitation, recombination, and dissociation of atoms, ions, and molecules by electron impact | |
| | *ION* | Ionization cross sections and rate coefficients caused by electron impact | 2862 |
| | *EXC* | Excitation cross sections and rate coefficients caused by electron impact | 716,452 |
| | *REC* | Recombination cross-sections and rate coefficients caused by electron impact | 27,532 |
| | *DIO* | Dissociation cross sections of molecules | 202 |
| *CHART* | | Cross-sections for charge transfer and ionization caused by heavy particle collision | 7618 |
| *AMOL* | | Cross sections and rate coefficients for electron–molecule collision processes | 3180 |
| *CMOL* | | Cross sections and rate coefficients for heavy particle–molecule collision processes | 2139 |
| *SPUTY* | | Sputtering yields for solids | 2349 |
| *BACKS* | | Energy and particle backscattering coefficients of light ions injected into surfaces | 485 |
| *ORNL* | | Bibliography on atomic collision collected at ORNL | 78,097 |

## 2. New System

### 2.1. Overview

Figure 1 shows an overview of the new system. The web interface was built using Ruby on Rails [7], which is a web application framework written in Ruby. The database management system was replaced by PostgreSQL, an open-source relational database system. Currently, ION data are copied into another database, which is dedicated to VAMDC node software [8], and are searchable from the VAMDC portal. Table 2 shows the main software components used in the previous and current systems.

The main reason for this update was that the OS and other software components had become outdated, and the security updates for the software were no longer being accepted. Therefore, we decided to upgrade the system to be compatible with the latest OS and other software. We also decided to use open-source software rather than proprietary software, because we could support it ourselves more easily. We chose Ruby on Rails as the web application framework, which can be used to develop a smoother and more interactive user interface. Now, a user can compose complex and multiple condition queries to search the data via a web browser. The user can also search from the periodic table interface for simplicity (Figure 2). The query results for the numeric data can be obtained in the form of tabular data, coefficients of the fitting curve, or graphic files (Figure 3).

### 2.2. VAMDC Interface

Through the VAMDC portal, a user can search simultaneously across a wide range of integrated databases from all over the world via a web browser. These database systems all comply with VAMDC standards, allowing the user to easily access their data. The standards define several parameters, including data models of atoms and molecules, protocols for queries and data retrieval, and so on. The portal is run via a node system. The VAMDC portal sends out queries and retrieves the results from the nodes by exchanging XML messages. To integrate the NIFS DB into the VAMDC node system (thereby making the NIFS DB a node), there were two possible courses of action, plan A and B (Figure 4). Plan A was to build a VAMDC interface on the present system. The web interface of the NIFS DB is implemented by Ruby on Rails, and it is possible to interpret XML using the Rails application to exchange messages between the NIFS DB and the portal. The advantage of this approach is that it is easy to maintain the consistency of the database. When the original data are updated, the changes in the NIFS DB are reflected in the VAMDC portal simultaneously. However, the necessary development required to implement Plan A would be complicated.

Plan B was to build another system dedicated only to the VAMDC portal using VAMDC node software—an application of Django, which is a python-based web application framework [9]. Adding little to no code, the system could easily become a node without modifying the original system. However, the database structure that the node software requires is different from that of the NIFS DB, and it would be necessary to synchronize data between the two databases.

After weighing these plans, we adopted Plan B, because the implementation of Plan A would have required huge modifications. Because updates of the NIFS DB are not frequent, real-time synchronization between the NIFS DB and the VAMDC database is not necessary, and it can be executed occasionally or manually. In the first step, only the ION table was copied into the VAMDC database, and currently, these are the only NIFS DB data available to be searched through the VAMDC portal.

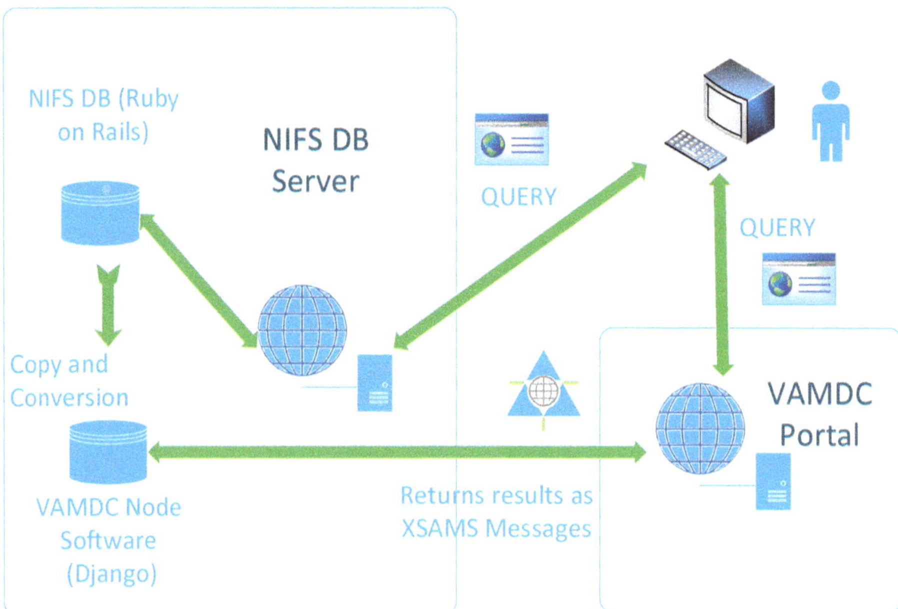


**Figure 1.** An overview of the new system. The contents of the NIFS DB are provided by the Ruby on Rails application. In order to communicate with the Virtual Atomic and Molecular Data Center (VAMDC) portal, VAMDC node software runs in the same server.

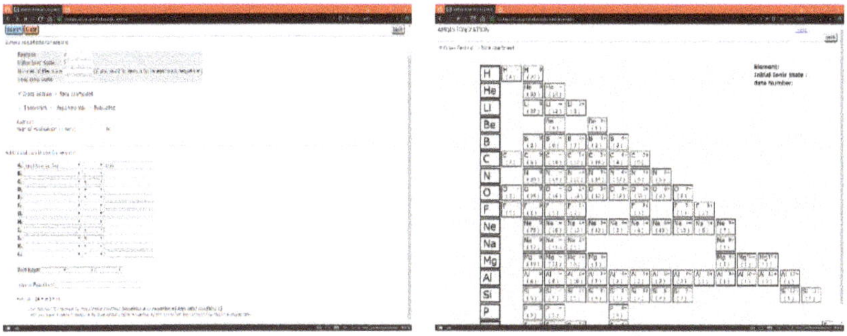

(a) (b)

**Figure 2.** The graphical user interface (GUI) of the NIFS DB. (**a**) The user can search data by constructing complex queries in the web browser. (**b**) For simplicity, the user can obtain the expected data of the atoms from the periodic table interface.

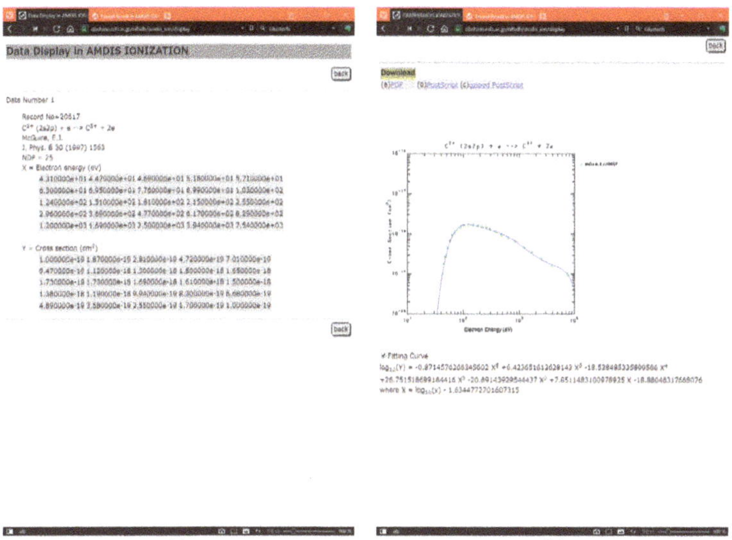

(a) (b)

**Figure 3.** The results of the search can be obtained in the form of (**a**) numerical data, (**b**) images (PostScript, PDF, PNG), or the coefficients of fitting curve.

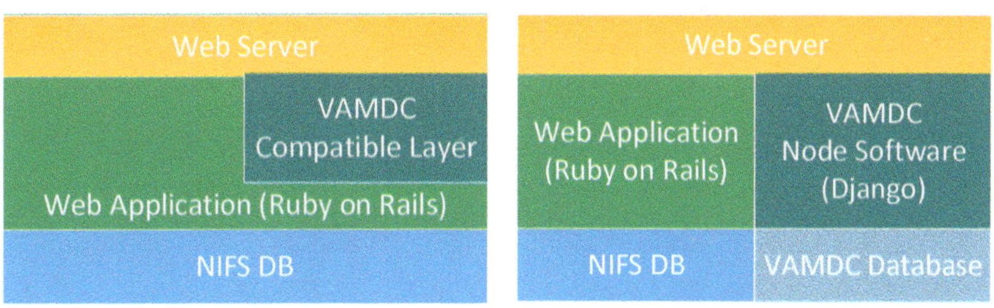


(a) (b)

**Figure 4.** The two possible integration plans. (**a**) Plan A was to implement the VAMDC standards in the NIFS DB by building node compatibility into the existing system. (**b**) Plan B was to build another NIFS DB system dedicated to the VAMDC portal, which was compatible with the VAMDC node software, Django.

**Table 2.** The major software components used for the old and current systems.

| | Old | Current |
|---|---|---|
| **OS** | RedHat Enterprise 5 | CentOS 7.5 |
| **RDB** | Oracle 8.5 | PostgreSQL 9.5 |
| **Web Application** | CGI | Ruby on Rails, JavaScript, Django |

## 3. Discussion

Because the initial design of the NIFS DB and the structure of the VAMDC database differed, the NIFS DB data could not be integrated into the VAMDC database in a straightforward way. For example, the original NIFS DB system compiles data of experiments and theories from published materials, and the database stores the tag of the original resource, to which the user can easily refer. In contrast, the VAMDC database treats atoms, molecules, transitions, and so on as objects. Thus, we decided to develop another NIFS DB system using the VAMDC node software. The conversion from the NIFS DB to the VAMDC database was executed by splitting the ION table into subparts (Figure 5).

In order to record as much published information as possible, the treatment of the data in the NIFS DB is neither consistent nor complete. Certain ION data have attributes of electron configuration, but some data do not. There are also some data that treat multiple reactions as a single reaction, such as the following:

$$\mathrm{A} + \mathrm{e}^- \rightarrow \Sigma\left\{\mathrm{A}^{n+} + (n+1)\,\mathrm{e}^-\right\}. \tag{1}$$

In such cases, only one set of numerical data for a single reaction is provided.

Currently, these data have been ignored and have not been copied into the VAMDC database. In order to access these data from the VAMDC portal, the modification of the VAMDC standards is required.

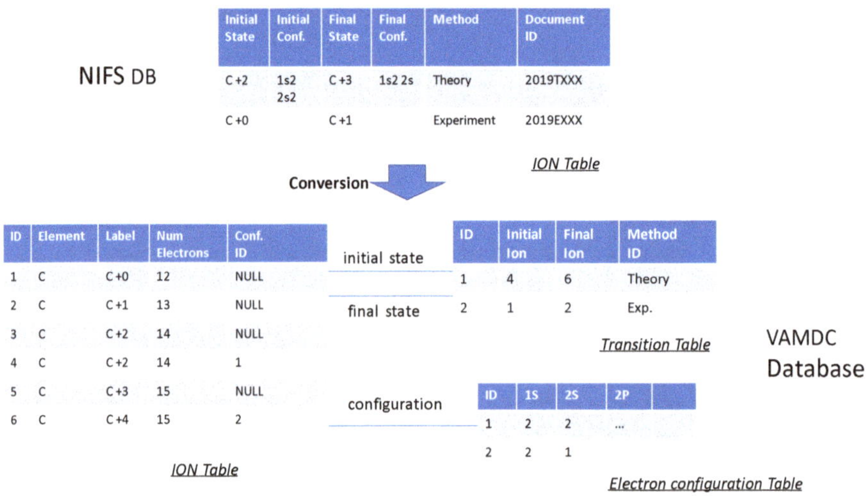


**Figure 5.** Conversion of the ION table from the NIFS DB to the VAMDC database. The original ION table in the NIFS DB is divided into three tables in the VAMDC database.

## 4. Conclusions

NIFS maintains the NIFS Atom and Molecular Database. Initially, the database was constructed to facilitate fusion plasma research, but it has also become useful in other areas of research, such as astrophysics. The data are now available online, and as a result of our recent updates, the data are now beginning to be available via the VAMDC portal. Currently, only the ION table has been made compatible with the VAMDC database, but work on making the other tables compatible is in progress.

**Author Contributions:** M.E. designed the new database system and managed the project. I.M. was the supervisor of the project. M.Y. and S.I. were the programmer. D.K. and M.K. helped us to build the database.

**Funding:** This work is supported by the NINS program of Promoting Research by Networking among Institutions (grant number 01411702).

**Acknowledgments:** We are very thankful to M.L. Dubernet and N. Moreau for their support in building the VAMDC interface. We would also like to express our gratitude to all those who gave us the opportunity to participate in this research.

**Conflicts of Interest:** The authors declare no conflict of interest.

## References

1. Takayanagi, K.; Suzuki, H. *Cross-sections for Atomic Processes Vol.1 (H) IPPJ-DT-48*; Institute of Plasma Physics, Nagoya University: Nagoya, Japan, 1975.
2. Takayanagi, K.; Suzuki, H. *Cross-sections for Atomic Processes Vol.2 (He) IPPJ-DT-50*; Institute of Plasma Physics, Nagoya University: Nagoya, Japan, 1976.
3. Kato, T.; Itikawa, Y.; Kanada, K.; Watanabe, R. Database and Retrieval Display System of Atomic Data for Fusion. *Phys. Sci.* **1981**, *23*, 198. [CrossRef]
4. Kato, T.; Tawara, H.; Matsunami, N.; Watanabe, R. Database and Retrieval System for Compiled Sputtering Data. *J. Nucl. Master.* **1984**, *128*, 1006–1009. [CrossRef]
5. Murakami, I.; Kato, D.; Kato, M.; Sakaue, H. Atomic and Molecular Databases and Data Evaluation Activities at the National Institute for Fusion Science. *Fusion Sci. Technol.* **2013**, *63*, 400–405. [CrossRef]
6. Dubernet, M.L.; Antony, B.K.; Ba, Y.A.; Bartschat, K.; Boudon, V.; Braams, B.J.; Chung, H.-K.; Daniel, F.; Delahaye, F.; del Zanna, G.; et al. The virtual atomic and molecular data center (VAMDC) consortium. *J. Phys. B* **2016**, *49*, 074003. [CrossRef]
7. Ruby on Rails Website. Available online: http://rubyonrails.org/ (accessed on 26 June 2019).
8. Documentation for the VAMDC Node Software. Available online: http://www.vamdc.eu/documents/nodesoftware/ (accessed on 26 June 2019).
9. Django Website. Available online: http://www.djangoproject.com/ (accessed on 26 June 2019).

MDPI
St. Alban-Anlage 66
4052 Basel
Switzerland
Tel. +41 61 683 77 34
Fax +41 61 302 89 18
www.mdpi.com

*Atoms* Editorial Office
E-mail: atoms@mdpi.com
www.mdpi.com/journal/atoms

www.ingramcontent.com/pod-product-compliance
Lightning Source LLC
Chambersburg PA
CBHW041758220926
51847CB00014B/248
* 9 7 8 3 0 3 9 2 8 0 8 8 9 *